Technical Writing
The Fundamentals

Third Edition

Wanda L. Worley
Rebecca Fitterling

Indiana University–
Purdue University Indianapolis

Kendall Hunt
publishing company

Kendall Hunt
publishing company

ISBN 978-0-7575-7927-1

Printed in the United States of America

10 9 8 7 6 5 4 3 2 1

Contents

9 WORKPLACE RESEARCH 191

APPENDIX A: RESOURCES 209

APPENDIX B: BRIEF HANDBOOK 231

INDEX 243

Acknowledgments

We would like to thank the following IUPUI students who have graciously and generously allowed us to use their work to illustrate our work. We hope their efforts will provide both help and inspiration to future students.

Paul Bailey
Jason Brugh
Anthony Costa
Eden Dorsett
Andrew Hoblitzell
Jacob Horn
Ken Jenkins
Stephanie Johnson
Chris Martin
Clint McCollum
Heather Pickering
Patrick Rasche
Morris Schefcik
Benjamin Scher
David Szpunar
Jim Wilson
Russell Washer

Many thanks to Katey Bertsch, acquisitions editor, and Amanda Smith, project coordinator, for her timely, responsive, and helpful efforts, and to all the staff at Kendall/Hunt Publishing for making this book possible.

~Wanda & Becky

CHAPTER 1

Introduction

Overview

As you make the transition from student to employee, you will find yourself challenged to make changes in your communication habits and styles. Although many of the communication concepts you learned in your college classes will apply to your workplace writing, you will need to adapt them to the peculiarities of the workplace and the particularities of the situation. This chapter presents a brief overview of what to expect from a class in technical writing, while focusing on its similarities with and differences from academic writing. It will also touch on the ethical concepts that underlie workplace technical communication and the influences of living and working in an international community.

Course Goals and Assumptions

The goal of *Technical Writing: The Fundamentals* is to provide you with a set of tools you will use to communicate in the workplace. We want to increase your awareness of the importance of good communication and your skill set for accomplishing that communication. As with many of your other classes, you may find the usefulness of this course in the application of what you learn, rather than in exact "how-to" formulas.

What are the challenges of preparing you for workplace writing? Many!

- People communicate differently.
- Technology is constantly evolving.
- Most businesses have a set of conventions to which they expect their employees to adhere.
- Often these writing conventions are not directly explained, so new employees learn by trial and error.
- What works today may not work tomorrow.
- People have numerous motivations and needs.

If we accept those statements as true, then we are challenged to present you with the skills that will help you adjust to and cope with this reality. To that end, we outline another set of assumptions:

- Understanding certain key concepts will aid in any kind of communication.
- Knowledge is not static; it changes and adjusts according to circumstances.

- Success requires flexibility.
- People need to learn and to apply what they have learned.
- The process is as important as the product.

What Is Technical Writing?

Fundamentally, *technical writing* refers to the kind of workplace communication that gets business done. It is practical, usable, accessible. It is concise. It is clear and accurate. It is formatted so that it can be read selectively. The subject matter and vocabulary tend to be specialized. Figures and tables are often used. In workplace writing, you are communicating to share knowledge that your colleagues do not (or may not) know.

If we accept the idea that the workplace is a busy, pressurized place with many activities going on simultaneously, then we need to accommodate that environment in the communication process. Therefore, certain common realities apply.

- Get to the point.
- Be accurate.
- Be complete.
- Be timely.
- Be clear and concise.
- Make the document easy to read.
- Format so information is easily accessible.

TECHNICAL WRITING VS. TECHNICAL COMMUNICATION

Very simply, technical writing is a part of technical communication. Technical communication includes text, visuals, oral presentations, and multimedia.

Academic Writing vs. Workplace Technical Writing

Academic writing, the kinds you have likely done, differs from workplace technical writing in purpose, audience, and approach, but does have many common features with workplace technical writing as well.

DIFFERENT PURPOSES AND AUDIENCES

You may wonder what challenges you will find in a technical communication class that differ from those in more academic settings. Fundamentally, academic writing has a specific purpose: to prove to the instructor that you have learned what he or she already knows. You write a paper or an essay (or take an essay test) in which you organize information around a major idea (thesis) and support that thesis with details.

Although you may be asked to synthesize information from multiple sources or to create your discussion around certain concepts, the receiver of your information—your instructor—is usually not new to the information. The basic purpose of academic writing, then, is to prove your knowledge to someone who is already an expert in that field.

Workplace writing differs on many levels from the academic model. The purpose of workplace writing is to get business done. You have information to share, or you need information, and that information has a specific business purpose. Two of the main purposes are to inform and to affect a change. Unlike your instructor, your reader in the workplace may be new to the information. The variables in the workplace context are numerous and changeable from situation to situation. Each writing endeavor addresses a specific set of circumstances.

Academic

Purpose =
To prove you know the material

Audience =
Instructor, other students

Workplace

Purpose =
To provide information to be used by the reader

Audience =
Supervisor, colleagues, technicians

SAME PRINCIPLES: DIFFERENT APPROACHES

Both academic and workplace communication have the same principle: to communicate information to readers. Any writing brings with it a variety of factors that you must consider. You are the keeper of information that you need to communicate.

In either context, as the writer you have the responsibility to accommodate your reader's (readers') needs and to make the communication as clear and as efficient as you can. In the workplace, you apply the principles of good writing that you have learned in academia for a very practical purpose.

Academic writing and workplace writing have the same principle—to communicate.

COMPARISON OF CONCEPTS

Although driven by different purposes, audiences, and conventions, academic writing and good workplace writing have many common features. Table 1.1 highlights those similarities and points out some of the subtle differences.

Table 1.1 Academic vs. Workplace Writing

Concept	Academic	Workplace
Main Idea	**Called a THESIS** Academic writing is built around a thesis, the guiding idea that shapes the argument and defines the supporting points. The academic piece is designed to prove the thesis.	**Called a PURPOSE** Workplace writing also has a clear main idea, though it may be simpler than an academic thesis. More often, workplace writing centers on the concept of *purpose;* what are you trying to accomplish by the writing?
Introduction	Academic writing uses the introduction to pull the reader into the piece, to establish the groundwork for the argument, to get the reader interested in the thesis, to establish the thesis. It is often written in the "funnel" format, starting with a general idea and moving slowly to the focused thesis.	Because business is so driven by time, the introductory material in a workplace report generally gets to the point much more quickly. Even when background information is necessary, the purpose of the report is clearly stated up front. Often the report begins with an orienting sentence for the reader ("Regarding your inquiry about . . . ," or "As we discussed last Monday . . .") and then moves directly to the information sought ("I have investigated three possible sites . . .").
Body	Academic writing relies primarily on the traditional verbal devices (in other words, consequently, for example, etc.) to make connections and tends to be more discursive.	Workplace writing, though certainly not devoid of verbal devices and transitions, also uses graphical cues: headings, subheadings, tables, charts, graphs. Even sentences take on a different look, as they are often reduced to lists and bulleted points for ease of reading. As a result, workplace writing has a much less text-filled look.
Conclusion	In academic writing, the writer often ties the final points together and reasserts the thesis.	In workplace writing, the writer makes sure the points are clear—restates them if necessary—or simply comes to a logical ending. Generally, because the workplace communication is directed at a certain person or persons, the writer addresses them directly, offering further explanation or availability. Since the purpose is to keep business moving forward, that goal is never far from the writer's mind.
Correctness	Academic papers require standard grammar, punctuation, spelling, and mechanics.	Like academic writing, workplace technical writing requires standard grammar, punctuation, spelling, and mechanics. You may be tempted to think that those details are less important in your workplace communication, but they are not. The communication you do in the workplace is a reflection of **your professionalism and attention to detail,** and though you may think you know who will be reading it, you can never be 100% sure. People forward messages or pull memos out of their files (even weeks or months later), and outside eyes may read—and evaluate. So the accuracy rule applies to workplace writing as well. Always put your best effort into your work.

Collaborative Writing

Along with communication skills, the ability to be an effective team player is one of the most important workplace skills an employee can have. Since many organizations have eliminated middle managers, a team approach is used more and more in the workplace. Being a team player often involves more than working on a project. A team project will likely lead to some sort of team-written document. So collaborative writing, although challenging, is common in the workplace, for several reasons. For one, rarely does one employee have all the expertise required to complete a project or to draft a document alone. In the scientific and technical fields, in particular, with their highly specialized and complex issues, it is routine to have several employees involved in drafting any major document. With networked computers and collaborative software, today's technology makes collaborative writing convenient for employees.

Multiple perspectives allow us to test our own knowledge against what others believe and, thus, we construct knowledge by interacting with others. These multiple perspectives, however, are what often make collaborative writing so challenging. We all have our own way of looking at things, our own style of expressing those views. Whether in the classroom or in the workplace, one of the biggest challenges to collaborative writing is melding all those perspectives, those unique styles of expression, into one, making the entire document flow as though one person wrote it. It's a good idea to have everyone involved read through the draft and make revision suggestions; however, one person must be responsible for that final revision phase to pull everything together. Although challenging, the pros of collaborative writing far outweigh the cons.

Ethics

- Respect
- Trust
- Honesty
- Integrity
- Responsibility
- Commitment
- Legitimacy
- Accuracy
- Fairness

These are all powerful words, commonly used in statements that companies publish about their "missions" or "values" or "codes of ethics." Think of ethics as the code of conduct of the organization, the structure of values upon which the policies, operations, products, and services of the company are based. Ethics are the infrastructure of the company, the foundation, and the frame. Without a strong infrastructure, the company stands like the little pig's house of straw—a determined wolf can blow the house down.

Of course, those words are only the labels put on ethics. The old cliché applies here: Actions speak louder than words. Respect, trust, and honesty in your business transactions affect the way employees treat each other, treat their customers, treat their vendors, treat their shareholders. Those values shape decisions. Those values shape communication in terms of openness, tone, word choice, and responsiveness. They serve to unify the work force.

Ethics extend to the larger picture as well. A company's approach to larger considerations is also shaped by its code of conduct. How do the company's actions affect the environment, for example? What is the company's role in terms of community involvement or education? How do the emerging issues of globalization, diversity, and technology affect the company's vision of itself and its ethical obligations?

As an employee of an organization, you represent that organization and its ethics in everything you do. Aware of that fact, you need to ensure that your communication practices uphold those values. You have control of the choices you make, and you will want to make them in the context of the company's code of ethics and your own.

International Issues

One final concept to discuss briefly is internationalization. Increasingly, even small U.S. firms find themselves with a customer and/or supplier base that extends well past the borders of the contiguous 48 states. In addition, their own employee pools come from diverse ethnic, religious, and cultural groups.

While the richness of this kind of diversity is hard to dispute, we would be naïve not to acknowledge that it also can make communication more challenging. Cultural biases influence whatever we do. For example, you will hear a lot in this book about the pace of the U.S. workplace and the assumptions we make about it: People are in a hurry; make information as concise as you can; take a polite tone but don't be overly chatty. Understand as you read this advice that it is not a pure, invested Truth, but rather a representation of how we see the realities of the U.S. workplace. Our advice does not hold true across the world, and, in fact, there are regional differences within the United States. We are verbalizing our assumptions but not idealizing them.

Communication is contextual.

You need to be responsible for what you say and do in the workplace. Be particularly sensitive if you receive an international assignment, either in a short-term or long-term capacity. Before you take an overseas trip, research the cultural conventions of the country you are visiting, and take the time to prepare yourself to act and react accordingly.

Chapter Summary

Clear technical writing incorporates many different elements. It is both similar to and different from the academic writing familiar to college students. Technical writing is rooted in a focused, purposeful way to keep business moving, while incorporating the principles of sound ethical judgment and respect for cultural differences.

End of Chapter Activities

ACTIVITY 1 Technical Writing vs. Other Writing

Bring to class two or three samples of technical writing and two or three samples of nontechnical writing. In groups of three to five, share your samples and discuss why you think they exemplify technical and nontechnical writing. As a group, brainstorm a list of characteristics that make a piece of writing technical. Be prepared to share your brainstorming list with the class.

ACTIVITY 2 Inventory List

Take about ten minutes and individually brainstorm how you would describe your positives and negatives as a technical writer. Join two or three other people in class and share your lists with each other. As a group, come up with your own definition of what a technical writer is. Be prepared to share your definition with the class.

ACTIVITY 3 Collaborative Writing

In small groups, go through the following scenarios and decide how you would respond to these situations if you were working on a group writing project.

1. Char and Danny are consistently late for class and often leave early, causing them to miss some group time on a regular basis. What do you do?
2. Adam usually never says a word in class or in group meetings. What he writes is short and concise, leaning toward curt. What do you do?
3. Emma is just the opposite. She doesn't stop talking. And, her writing is verbose. What do you do?
4. Gloria thinks she knows it all. She dominates group meetings and thinks her way of expressing ideas is far superior to everyone else's. What do you do?
5. Blake is a workaholic and wants to do all the work. He's highly capable and writes well. What do you do?

ACTIVITY 4 Ethical Assessment

Pick a company with which you have some familiarity—it could be the company for which you work or would like to work, a well-known local company, or even a nationally known company. Now, look on the company's Web page and see what it says about itself: its mission statement and core values, its commitment to customers, its attitude toward its employees. What are the ethical principles it espouses? As a customer or an employee of that company, how would you expect to be treated? Does the company offer evidence of its "walking the walk" as well as "talking the talk"?

ACTIVITY 5 Action Plan for Improvement

What areas of your own writing would you like to improve? Complete the following worksheet and submit it to your instructor.

Action Plan for Improvement

Define ⇒ Plan ⇒ Act

Take a few minutes to think about the qualities and skills you would like to have as a writer. Then answer the following questions as completely as you can at this time. After you leave this class, continue to refine your action plan until you have it the way you want it. Revisit your action plan weekly, if not daily. Ask yourself these questions: Are my goals still the same? Am I working my plan? What changes do I need to make to the plan?

1. *Define*
   A. What attributes and qualities do you most want to develop as a good writer (for example, professionalism, accuracy)?

   B. Describe your best qualities and skills as a writer.

   C. What skills do you feel least comfortable with or feel you need to improve?

2. *Plan*
   A. Based on what you just wrote, write down at least three goals you'd like to accomplish in the next six months. Remember, goals need to be specific, measurable, realistic, and action-oriented.

      1.

      2.

      3.

   B. Put the three goals you just wrote in order of priority.

      1) 2) 3)

C. List the steps required to complete each goal.

- Steps to accomplish goal 1:

- Steps to accomplish goal 2:

- Steps to accomplish goal 3:

D. Now determine a deadline to accomplish each of the steps you just wrote down.

Steps	**Deadline**

3. *Act*
Start accomplishing your goals one day at a time.

CHAPTER 2

Writing Process and Strategies

Overview

Chapter 2 takes you through the steps and stages that most writers go through to produce a successful piece of technical writing. While certainly not everyone follows the exact same steps in the exact same order, this chapter offers a look at the kinds of decisions and processes typically found in the workplace.

Writing Process

Planning ➤ Designing ➤ Writing ➤ Revising ➤ Getting Feedback ➤ Revising Again ➤ Reflecting

PLANNING

Wouldn't it be wonderful if, when given a writing task, all we had to do was sit at our computer and pound it out? In the blink of an eye, job done, document produced, satisfaction guaranteed.

That scenario doesn't happen very often. Each writing situation has its own variables, its own quirks, its own politics to attend to. Nonetheless, techniques exist that can help you gain some control of the process and ultimately help you expedite it. One of the goals of this book is to provide exercises and tools that will sharpen your writing skills.

The first step is to understand (and accept) that writing is a process. This process involves planning, and planning involves making several decisions. For example, you have choices regarding organization, tone, detail, word choice, approach, design. You have explicit and implicit agendas to satisfy. You have an audience to address, which may be one person or multiple people. You have a purpose to accomplish.

Fortunately, you come fully equipped with a brain, so those decisions need not overwhelm you. The wisest course to take is to approach the task systematically, starting with some preparatory work. You can describe this early work in any number of ways—brainstorming, prewriting, planning, drafting, data collecting—you name it. The crucial concept here is that this idea-generation helps you collect information, some of which you will use and some of which you may discard. It starts you on the road to your final product.

Planning a short document usually isn't as daunting as planning a long document. Facing the task of writing a long report can be overwhelming, even paralyzing. A little time spent planning the direction of the report can ease some of the anxiety. Also, the time and work you spend planning will result in less time writing and revising. Common sense, though, tells us that each document we write is unique and, thus, each planning stage is unique. However, the more practice you have, the more efficient and effective you become. Over time, you may even do the majority of your planning in your head. You must find the process that works best for you, but no matter the process, you need to answer these questions:

1. Who is your reader?
2. What is your purpose in writing the document?
3. What background information do you need to consider (and possibly share in the document)?
4. How will you organize the information?
5. How will you gather information needed?
6. What tone do you want to convey?

Audience/Reader

The first crucial variable to consider is audience/reader. The roles of student and teacher generally put a fence around the communication, and that fence is usually fixed. In the workplace, you will have many relationships, each dictated by the person's role and function in the business. You will communicate with many different people in many different capacities, and you will need to be sensitive to each person's role. Titles often give clues to roles, so pay attention to the titles people carry. In addition, keep the general categories in Table 2.1 in mind.

Think like your reader.

Clearly, your audience's expertise and functions will influence your decisions for presenting the information you are communicating. Frequently, one key complication is that you must address more than one person and more than one functional category. Perhaps you know that before your manager can be persuaded to make a change, he will consult with the experts in the company for their input. Therefore, you not only have to fashion your communication for your manager, but you also must incorporate enough information for the experts so that they feel comfortable endorsing the change. In cases like these, focus the attention on the **primary audience,** but be sure you provide enough information for the **secondary audience** as well. The brainstorming questions in the following box will help you think about your reader.

BRAINSTORMING QUESTIONS—READER

- *To whom* are you writing this report? In other words, who is your reader? Give your reader a name.
- *Is* the reader a manager, an expert, a technician, or a general reader?
- *What* is your primary reader's relationship (within company) to you—boss, colleague, etc.?
- *What* is your primary reader's age and background?
- *Why* will your reader be interested in your report?
- *What* does your reader already know about the topic? Need to know about the topic?
- *What* reaction do you anticipate from this person when the details of the situation are known? Would s/he classify the situation as high importance, moderate importance, low importance?
- In addition to your primary reader, *who* in the company may need to read the report?

Table 2.1 Reader Roles and Functions

Category	Function	Typical Titles
Management	Decision maker. Often the person who assigns the responsibilities, makes sure the project and people are staying on target, reports to upper management. Very often in charge of the budget, making the decisions associated with money. Many personnel responsibilities, including coaching, performance evaluations, hiring and firing. (The higher up you go, the more financial responsibility you get, and the less hands-on personnel work you do.)	Manager General manager Higher level: assistant director, associate director, director Higher yet: assistant vice-president, associate vice-president, vice-president Higher yet: executive vice-president, chief financial officer (CFO), chief operating officer (COO), treasurer, managing director Highest: president, chief executive officer (CEO)
Expert	Specialized knowledge. Often the go-to person when detailed explanation or analysis is needed. Focus is often on knowledge more than on people. Most often used as resource rather than as decision maker.	Specialty titles: accountant, engineer, systems engineer, project coordinator, human resources specialist Generically: specialist, coordinator, consultant, advisor, subject matter expert (SME), analyst, administrator
Technician	Hands-on person. Makes systems, machinery, processes work. Often needs specialized training in order for that to happen. May be used for troubleshooting and maintenance.	Technician, mechanic, operator, administrator
General reader	Interested outsider. May have no direct role within the company, but has an interest in what the company is doing.	Shareholder, community activist, legislator

Purpose

Another crucial variable that you need to consider in making your writing choices is the purpose for sending the communication. While your academic purpose may be simply to get the paper written and submitted to the instructor to prove yourself worthy of a certain grade, workplace communication frequently is a bit more complicated. In many respects, purpose cannot be separated from the audience because different roles have different needs. Thus, you need to know not only why you are communicating, but also to whom and, therefore, in what way.

To Inform = Explain, clarify, instruct

To Persuade = Recommend, convince, influence

Think of purpose in terms of a verb. The standard verbs are **to inform** (what day is the meeting, who is on the committee, how should the vacation request be completed) and **to persuade** (change the vacation policy, create a new position, purchase new equipment). Apply these concepts to the person you are communicating with: What informational needs does this particular person/position have, and how can you best provide them? Or, how can you best persuade this particular person/position to accept your ideas?

Answer the brainstorming questions in the following box when thinking about the purpose of your e-mail, report, or document.

BRAINSTORMING QUESTIONS—PURPOSE

- *Why* are you writing this document? Are you informing, recommending, warning, persuading?
- *Who* initiated the communication? You or someone else?
- *When* do you need to get it done?

If your supervisor initiates the communication, the supervisor may already be convinced, so your purpose may tend to be more informative than persuasive. However, if you initiate the communication, you may have to be more persuasive in your approach.

Background

When planning your report, you will also want to think about any background information that influences your message and your e-mail, report, or document. Consider the questions in the following box as you think about background.

BRAINSTORMING QUESTIONS—BACKGROUND

- *What* is the subject of this document?
- *What* situation is prompting the communication? Has it been going on for a long time, or is it a recent development?

- If a problem, *what* caused the problem? *What* effects is this problem causing?
- *Who* is involved?
- *Why* is it important to look at this situation now?
- *When* does something need to be done?
- *How* do you explain it?
- *How* would you classify the situation: as high importance, moderate importance, or low importance?

Organization

In addition to thinking about your reader, the purpose, and important background information, you'll also want to plan the document's organization. As you think about organization, consider the questions in the following box.

BRAINSTORMING QUESTIONS—ORGANIZATION

- *How* should you present this information?
- *What* are the main and supporting points you want to make?
- *How much* detail do you need to provide to your reader?
- *How* would graphics help the reader comprehend what you are saying?

Research

Planning involves thinking about how you will gather the information and data for the e-mail, report, or document. The following box contains questions to consider.

BRAINSTORMING QUESTIONS—RESEARCH

- *What* have you done to gather information about the topic?
- *To whom* have you spoken? Have you consulted any sources, talked to any subject matter experts?
- *What* hard data can you report, if any?
- *How detailed* do you need to be in reporting your research?

Tone and Word Choice

Finally, you must consider tone and word choice as you plan your document. In keeping with what Chapter 1 advises, you must first consider the businesslike nature of the workplace report, and that means keeping a professional tone in your writing. You are writing to a reader, and you must respect both that reader's business role and relationship to you as well as his or her own personality. Most of us prefer to be treated politely and respectfully. If we operate with the understanding that the ethical stance our company takes is to treat people openly, politely, honestly, and fairly, then we communicate with our co-workers and clients with those qualities in mind. We avoid accusation, sarcasm, condescension, and negativity; we strive for fairness, clarity, temperance, and positivism. We are aware that certain hot words may trigger negative responses in our reader. We must keep in mind that our purpose is to move the business forward in the best way possible.

Choosing positive, accurate, and clear words is only part of the overall challenge of word choice. You also have choices to make in terms of industry terms and jargon, level of technical detail, and company shortcut phrases. Understanding your reader's role in this situation and his experience with the issue helps guide those choices. As you get to know the people in your company, especially those with whom you communicate frequently, the choices become second nature; you will understand, for example, what terms need to be defined for a certain person or what words you should avoid with certain others. Even when the choices seem to be less daunting, however, you should always keep in mind the general principle of making your communication the best representation of yourself as you can. Always be mindful of tone and word choice. Refer to Chapter 8 for further discussion on tone and language use. Also, check out the following box for brainstorming questions.

BRAINSTORMING QUESTIONS—TONE AND WORD CHOICE

- *What* overall impression do you want to leave with your reader? (Put a word on it: serious, informed, concerned, or convinced, to name a few.)
- *What* words could you use to convey that impression?
- *What* words do you need to avoid?

DESIGNING

Once you have brainstormed answers to these kinds of content questions, you have content to work with. Now all you have to do is decide how you want to present the information. We call this the **design stage:** You begin with your basic memo or e-mail or report format and make decisions that will best represent your ideas. In this stage, you'll develop your direction, a roadmap of sorts, intended to guide your reader through your communication. Some people may call the design plan an outline. See the following box for questions to consider as you design your document.

BRAINSTORMING QUESTIONS—DESIGN PLAN

- *What* headings will you need for your report? Will you have subheadings under the headings?
- *What* graphics have you decided to include? *Where* will you put them?
- *In what order* will you offer your supporting points?
- *How* will you connect the points?
- *How much* background will your memo contain?

WRITING

Finally, having made your preliminary decisions, you begin to write. Now it is time to **consolidate** and **synthesize** the information you have generated in the planning and designing stages. Use your brainstorming material and your design plan to put together the first draft of your document.

Now go for a walk or fix yourself a sandwich or hunt down bad guys on the Xbox for a few minutes or . . . you get the idea. Do something that takes your mind off the writing task. Taking a break from your draft helps you see it fresh when you return to revise.

REVISING

Wouldn't we all love to be perfect communicators the first time, never needing to revise even the smallest word? The reality is that we aren't perfect communicators, and even when we play both creator and editor as we write, we still miss things.

Revision has many different layers: You can revise for ideas, for organization, for sentence structure, for word choice, for tone, and on and on. Ideally, when you return to your document, you will read it with fresh eyes. Ask yourself if you have accomplished your objectives. Consider the following questions.

REVISION QUESTIONS

- Have you addressed your audience appropriately?
- Have you conveyed the right tone?
- Is the memo organized well so that the reader can easily follow your points?
- Did you provide enough detail?
- Is your purpose clear?
- Do the sentences read easily; are they clear and grammatically accurate?

Generally, your fresh reading will at least raise some questions in your mind. You may be unsure about your success at accomplishing your goals. You may feel inadequate in making that judgment about your own writing. In either case, you have reached the stage where you will want to consult a trusted colleague.

GETTING FEEDBACK

Peer Review

You are probably familiar with the peer review process, having undoubtedly encountered it in composition classes. Some students may react negatively to peer review because they feel unqualified to give input and are distrustful of the input they receive. Writers feel very vulnerable when someone is reading their writing, and often that vulnerability causes discomfort and tension.

Yet, you need to understand that peer review happens every day in the workplace; it is not just an academic exercise. Let's say your manager has asked you for some information, and you have done the research and written a report. You are concerned that the manager is going to react negatively to your research, and you are trying to present it in as neutral a way as possible. Rather than pushing the "send" button on your computer, ask a fellow worker to read it first. You trust the advice because you trust your co-worker, based on your experiences with him and your knowledge of his abilities. And the result will probably be a better response from your manager. You have lessened your risk!

That kind of trust, of course, comes with time. Though you may not initially have that kind of trust in your classmates, with experience and practice, it will come. You can help that process along by letting the reviewer know your concerns. Look at these examples:

"I don't want this memo to sound whiny; do you think my tone is all right?"
"I want to convince Sue to change vendors; do you think I made my case? What else should I say?"
"I know what I'm trying to say in the last sentence, but I don't think it's very clear. Do you get it?"
"Can you think of a better word to use in that first sentence?"

If you provide your peer reviewer with guidance about your needs, she will feel more comfortable in supplying you with reactions.

Although, ultimately, you as the writer are responsible for directing the peer review, the guide sheets we have designed will help focus the review. See Appendix A. Keep in mind, the goal here is to receive productive, reader-based input on your writing. You want to know how the potential receiver of your message is reacting to your communication. See Figures 2.1–2.5 at the end of the chapter for sample peer reviews.

One word to avoid in the peer review process is *critique*, because it carries negative associations. Think of peer review as the reader's *response* to your communication, rather than a *judgment*. Peer reviewers need to take responsibility for their response (e.g., "I'm not sure I understand your point here"). "Your point isn't clear here" is a judgment statement. You as the writer are more likely to accept the first statement than the second one.

The questions you see on the peer review worksheets in Appendix A state the elements of an efficient, clear document. Consider the following underlying concepts.

UNDERLYING CONCEPTS

Clarity

- After reading the document, do you think the intended reader will understand the purpose?
- Does the subject line serve that purpose?
- Does the writer have a good guiding verb in the introduction (examples: compare, recommend, convince)?
- Are the sentences concise and easy to follow?

Organization

- Has the writer put ideas in a logical, easy-to-follow sequence? In other words, has the writer chunked the information in ways that make sense to the reader and used headings and subheadings to identify those chunks?
- Are the chunks logical and logically ordered?
- Do they serve as guideposts to the reader?
- Has the writer used transitional phrases such as *in addition to*, *instead of*, *as a result of*, or words such as *first*, *second*, *third*?
- Does each chunk of information stand on its own in terms of having a clear introductory sentence, supporting ideas, guiding language?

Complete and Accurate

- Is the information complete and accurate?
- Has the writer provided the reader with just the right amount of supporting facts, details, data to prove that he knows what he is talking about and that the reader's needs and interests are a priority?
- Has the writer done the research to make sure those "facts" are facts and that the details and data provided are accurate?

Error-Free

- Has the writer revised and edited the document so that it uses the standard conventions of grammar, punctuation, and mechanics?
- Has the writer proofread to make it error-free?

Usability Testing

If you have written a set of instructions, now's the time to schedule usability testing of the instructions. See Chapter 6 for more information on usability testing.

Editing vs. Proofreading

Two other words frequently attached to the peer review process and not infrequently used synonymously are *editing* and *proofreading*.

There is, in fact, a difference between these words. Editing looks at the *content* and *presentation* of the report: Is the tone appropriate; are the points clear and understandable; has the writer explained the idea clearly; does the report need more detail; will the manager take offense at something in the message? Proofreading, by contrast, has to do with making sure that the *mechanics* of the paper are accurate: Are the apostrophes placed correctly; has the writer spelled all the words correctly; do you see any comma splices?

Although both of these concepts contribute to the overall effectiveness of the communication, editing probably should take at least chronological precedence over proofreading—you probably are in jeopardy if your perfectly spelled and formatted memo contains provocative language! However, your perfectly crafted message may lose its effectiveness with sloppy misspellings and misplaced commas!

REVISING AGAIN

Now that you have feedback from a reader or readers, you want to revisit your document again. Depending on the review process in your workplace, you may or may not have to accept the revision suggestions you receive. In the classroom, however, you as the writer will decide whether to take the advice of your reviewers. You may say to yourself, "I see her point, but I don't agree with it." You do not change your document. Or, you may say, "Duh . . . how could I have missed that?" You do change your document. Whatever your reaction, at least look at the feedback and consider the merit of the suggestion.

As instructors, we have heard the following comment more than once: "Why should I take another student's advice? He doesn't know any more than I do." We suggest that all of us have worthy advice to give to writers. We've been reading since a very early age. We know what "sounds right." We know if the writer has written the document in a way that we understand. We know if the writer has convinced us that she has a complete (or accurate) picture of the situation. Sure, we may not be the grammar guru of the class or a punctuation expert, but we all have valuable feedback we can offer a writer.

REFLECTING

By now you've probably gotten into the rhythm of the composing process—from the early stages of picking a topic and thinking about the context; to collecting the information and putting it into a format that suits your purpose and audience; to asking for input from others and doing revisions; to proofreading the copy for clean grammar, punctuation, and presentation; and finally to reflecting on the process. You may be wondering about this last step, the reflection. What is its purpose in this whole scheme of things? What does reflection have to do with workplace writing?

Reflective writing is writing that asks the writer to look back and articulate what has transpired from the start to the finish of the writing experience. The principle behind reflective writing is that putting something down in words makes it more real to you. Only when you know what you have done can you make needed changes. You will be able either to duplicate or to revise your strategy according to how well what you did before worked for you. Instead of operating in a random way, you are learning to identify your own process, your own strengths and weaknesses:

- What part of the process did you find most challenging—planning, writing, revising, editing?
- Does writing a brainstorming list work?
- When did you make formatting decisions? Would you change anything the next time?
- What kinds of comments did you get from the peer reviewers? Have they told you something to keep in mind when you write your next draft?

Articulating your own process, then, will help in a very practical way by allowing you to control your own activities in future writing tasks. You will probably learn to budget your time and energy more efficiently. That means, of course, that the time spent on unproductive activities will lessen, and that you will be able to use your time wisely and well—certainly a workplace skill. For example, if you know that you use a first draft in lieu of an outline, then skip the outline and write the draft. On the other hand, if you find that an outline helps you visualize your final communication and expedites your process, use an outline! Reflective writing has very practical effects.

Of course, this concrete application is only the start. Once you have mastered the practical considerations, you will find yourself moving up the abstraction ladder to the next step, where you will start understanding your writing process in terms of other characteristics:

- When you collaborate, for example, do you tend to hold back comments or burst forth with them?
- Do you prefer to work alone or in community?
- Do you find certain cognitive or creative activities easier than others?
- Does that insight track with other things you know about yourself?

- How have you changed over the years? Is what you are learning in your communication class affecting other parts of your life?
- Are you observing similar changes in your classmates?
- How will these insights help you as you enter the work force?

Finally, your reflection will lead to thinking about bigger issues. How does what you are learning in the classroom and at the university inform what you know about the "real world"?

- What are you observing about the human dynamics around you?
- Have you been affected by the diverse university environment?
- Have old predispositions or judgments changed in the process?
- How do your communication habits relate to the ethical choices that you make every day?

In other words, the reflective process is moving you toward a better self-understanding. It strives to encourage your intellectual growth and your critical thinking, to use yourself as the subject of study. These ideas are the "larger" ideas of education. They may be articulated as principles of undergraduate learning in some universities. In engineering and technology schools, ABET, the accreditation organization, calls them program outcomes. In your own lives, they are what make you valuable employees and citizens.

See Figure 3.3 for a sample course reflective memo.

Sample Peer Reviews

The following pages contain samples of productive and helpful peer reviews (Figures 2.1–2.5). Included are both questions from the peer review worksheets and from the Track Changes version of the document.

These samples came from former students. Notice that the first student actually asked her reviewer specific questions that she had on her draft, helping him know in what areas she most needed advice. Note, also, that the students used a lot of "I" statements to provide the writers with their honest, reader-based feedback. They also made specific, positive comments where appropriate. You will notice the samples are only part of the whole document.

Peer Review Worksheet

1. **Opening Sentences**

   The opening sentences under each heading should guide you to know what you will be doing and why.

   __Y _x_N **Do the opening sentences tell you what you will be doing and why? Explain.** Under Utilizing Top Producer, I think a couple of introductory sentences would be helpful for this step. There are a lot of things going on here, and it will help prevent having a header leading into a header.

2. **Visuals** (Visuals are essential in a manual.)

   _x_Y __N **Do the visuals fit the text? Explain.** The visuals are very helpful in explaining what is going on, and they fit the text well.

   _x_Y _x_N **Are the visuals placed appropriately in the text? Explain.** For the most part, they are placed right where they need to be, but, there is one figure that I think needs to be placed in the manual a couple of times. This is Figure 3.2. This figure is referenced through many steps, but, you have to flip forward 2 pages in order to see it. I think it needs to be on multiple pages.

   _x_Y __N **Are the visuals labeled (e.g., Figure 1: Side View of . . .), documented (e.g., Source: Thomas, p. 30), and referred to in the text (e.g., See Figure 1.) Explain.** All the figures are labeled and documented with the appropriate information.

Figure 2.1
Sample Responses on Partial Peer Review Worksheet

Heather Pickering: Writer—Manual, Worksheet Responses
Ben Scher: Peer Reviewer

How to Use Top Producer

Welcome

Congratulations on purchasing a subscription to Top Producer 7i. You will find that Top Producer is a highly effective and efficient way to manage all of your real estate needs. Top Producer can look overwhelming at first, especially to those without a lot of computer knowledge, but you will quickly find that it is very easy to learn and use. Just be sure to be patient and kind to yourself; you will not remember it all in one day, and that is okay! Also remember that your co-workers have been here before, too, and our doors are always open to help answer your questions.

Getting Started

These instructions were written for users running Top Producer on Windows XP. If you are not running Windows XP, the instructions may be slightly different; please contact the IT department for questions regarding your specific computer.

Materials Needed

- A subscription to Top Producer 7i (agent license)
- A computer with Internet Access
- Contact information to enter into Top Producer

Estimated Time of Completion

The time of completion depends on how many contacts you intend to enter. It takes about 15–20 minutes to login, enter one new contact, and log out for the first time. This time will decrease as you become more familiar with Top Producer.

Utilizing Top Producer

1. Logging In

1.1 Do you have a Top Producer Icon on your Desktop?

Login to TOP PRODUCER 7i.url

A. If *yes*, Double click the icon using the left mouse button.

Comment [cnc14]: I am assuming that "our doors" are the manufacture of Top Producer, or is it someone else? BAS

Comment [MSOffice15]:
Ben,
Can you please let me know if there are items you think should be defined as you go through this? I really did not feel that any of the terms needed to be defined further than they already were.
Thank you very much,
Heather

Comment [CNC16]: Heather, overall, it is pretty cut and dry what it is you are talking about. The hardest thing for me, a user with no previous knowledge of this system, is that I have to flip back and forth between instructions and pictures. Unfortunately, I learn a lot from seeing, so I don't know if it would benefit to have the same figure put in the work more than once. BAS

Comment [CNC17]: Is this a program designed for everyday use, or does the end user get in, do their work, then get right back out? BAS

Comment [MSOffice18]:
Ben,
Another thing I am concerned about is having a heading leading into another heading. Do you have any suggestions for this situation?
Thanks!
Heather

Comment [CNC19]: Heather, one possible way to prevent a header leading into another header is to maybe put a small paragraph here that summarizes what the user is going to do in the following steps. This way, you break up going from header to header, and the reader has a little bit of understanding which direction they are going before they jump right in. BAS

Figure 2.2
Sample Partial Peer Review Showing Track Changes

Heather Pickering: Writer—Manual, Track Changes
Ben Scher: Peer Reviewer

Comment [CNC20]: Heather, what do you do if you don't have a username or password? Where do you go to get one? BAS

B. If *no*,
- Open the Internet browser.
  Note: Please consult IT if you have problems doing this.
- Type https://www.topproduceronline.com/default.htm into the Internet browser address bar.
  Note: The Web address must be typed exactly as shown above.
- Press the "Go" arrow/button in order to be redirected to the Top Producer login page.

1.2 Type your username into the "Username" box.
1.3 Select the method you wish to use for searching.
Note: Last name is usually the best method to use when checking for redundant entries.
1.4 Type the search information into the "Search for" box.
Note: When searching by last name, an example would be typing in "Smith" (without the quotations) for a contact named Joe Smith.
1.5 Click the "Start Search" button at the bottom of the page.
1.6 Was a contact found?

A. If *yes*, does the contact(s) match the contact you were going to enter?
- If *yes*, Click the "Close" button at the bottom of the page, and proceed to "Logging Out."
- If *no*, Click the "Close" button at the bottom of the page, and proceed to "Entering a New Contact."

B. If *no*, a box will pop up that says "No matches found."
- Click "OK."
- Proceed to "Entering a New Contact."

Comment [CNC21]: I am a little confused here, if you search for a contact, and they are in the system, why would you then log out of the program? BAS

Figure 2.3
Sample Responses on Partial Peer Review Worksheet

Ben Scher: Writer—Recommendation, Worksheet Responses
Heather Pickering: Peer Reviewer

Peer Review Worksheet

1. **What background information does the writer give to help you understand the situation?** The current 401(k) plan is too complex when it comes to selecting the proper mixture of funds, not straightforward enough that employees can easily understand what the plan offers, and too expensive for the individual employees.
2. **Is the situation thoroughly investigated and explained? Explain.** You did a great job of writing this recommendation, but how many employees does this problem affect? What will be the cost to the company to change 401(k) plans? How quickly could the new plan be implemented? Is the company obligated to stay with the current plan for a certain amount of time (i.e., does the company have any sort of contract with the current plan . . . I really don't know if that is how these things work)? Has anyone else tried to change the current plan? If so, why did that person's attempts fail? In regards to your research, how did you narrow the mutual companies down to these three companies? How did you get your information about each company?
3. **What sources does the writer use to research the situation?** I am not sure. See my comment about "how did you get your information about each company." Make sure to cite your sources, too, if it came from a Web page. If you got your information from talking to company representatives, it might not be bad to cite the conversation (especially date) so that you could go back to them and say, "on xyz date, you said the company could get $X discount" or whatnot.
4. **What are the results of the writer's investigation?** Convenient Mutual Group is the best choice for the company.
5. **What do you still need to know to fully understand the situation, why the writer is comparing these options?** How is this change beneficial to the company? You did a great job of explaining how it is beneficial to the employees, but kind of left out the aspect of how it is beneficial to the company.

. . . The three mutual fund companies that were researched based on the above criteria were:

1. Express Mutual Funds
2. Convenient Mutual Group
3. Future Funds

> Comment [MSOffice22]: This part may be redundant; you have already indicated that the options were compared on the criteria above. (Just throwing that out there because I tend to have problems with being too redundant.) HPG

All three companies have a respectable history as well as assets in the billions of dollars.

None of the companies have been involved in scandal, corruption, or fraud. All their holdings are listed on the New York Stock Exchange.

> Comment [MSOffice23]: I'm not sure you need to change this, but remember that the book cautioned against using clichés. HPG

Being an investor, it is very important to understand and see how mutual fund companies are handling my investment money. As with any consumer, it is important to get "the most bang for the buck" that is possible. In order to do this, I created a three column log which was used to compare my three criteria against my three chosen mutual fund companies. The results of these findings are listed in Table 1.

> Comment [MSOffice24]: You have 4 criteria, not 3. HPG

> Comment [MSOffice25]: Don't forget to label and document your Table. HPG

A = Express Mutual Funds B = Convenient Mutual Group C = Future Funds

Table 1 Mutual Fund Comparison

Ideal (criteria)	Real (evidence)			Judgment (difference)		
	A	B	C	A	B	C
Expense ratio	0.21%	0.76%	0.79%	+	+/-	-
Past performance	12.41%	16.20%	1.80%	+/-	+	-
Web site usability	Good	Great	Average	+/-	+	-
Age of fund (birthday of fund)	10/27/03	5/31/05	6/1/06	+	+/-	-

By Benjamin Scher

Based on my research and the information posted above, I recommend going with Group B, the Convenient Mutual Group. Their expense ratio was higher than Group A, but they more than made up for that in performance differential. Convenient Mutual

> Comment [MSOffice26]: Maybe say "presented in Table 1" HPG

> Comment [MSOffice27]: I would say "Convenient Mutual Group's expense ratio. . . ." HPG

Figure 2.4

Sample Partial Peer Review Showing Track Changes

Ben Scher: Writer—Recommendation, Track Changes
Heather Pickering: Peer Reviewer

Group has a Web site that is extremely easy to navigate and understand, as are all of the online forms that may need to be filled out. Finally, even though the fund from Convenient Mutual Group has only been around since May of 2005, Convenient Mutual Group has been around for decades, solidifying its reputation in the mutual fund business.

Comment [MSOffice28]: This may be a run-on. Try using a semi-colon to break this up. Maybe. . . "decades; this solidifies its reputation in the mutual fund business." or something like that; I just think it should be broken down a bit. HPG

Conclusion

Comment [MSOffice29]: Consider changing this heading to "Recommendation" HPG

With the hustle and bustle of everyday life, it is not hard to understand why employees do not register for the company's 401(k) plan. With the complexity of how to invest properly, a common fear among employees about losing money, and difficult times employees face with life's everyday challenges, I think it is time for a change. This change is directed toward our company retirement program. I think it is time to cancel the current plan, and move on to better things.

My recommendation is to move in the direction towards Target Retirement accounts, specifically from the Convenient Mutual Fund Group. Not only will this change take the complexities out of investing, it will increase the employee's confidence level about their decision making regarding their future. This will prove to be beneficial not only to the company, but to each individual employee as well. If you have any questions, or if there is any further information I can provide, please do not hesitate to contact me.

Comment [MSOffice30]: Good point! HPG

Comment [MSOffice32]: I just threw this in so you would not forget to add a "cooperative" sentence. ☺HPG

Comment [MSOffice31]: How is it beneficial to the company? HPG

Figure 2.4

Sample Partial Peer Review Showing Track Changes *(continued)*

Figure 2.5
Sample Student Peer Review (Jacob Horn, Peer Reviewer)

To: Tom Heraly
From: Paul Bailey
Date: October 11, 2009
Subject: Problem Analysis on Air Compressor Problems

Summary
This report is to inform you about the problems that have come up with the main air compressor controlling all air lines to the building. The problem that we are having is that the water pump that keeps the water out of the air lines has gone bad. This is affecting most of the building causing various problems in each department. If not this problem isn't fixed soon, there will be bigger problems at hand which will cause us to shut down production until it is fixed. We need to replace the pump and figure out why the pump has gone out so it will not happen again.

> Comment [MSOffice33]: I would suggest rewording this because you identify that there is a problem with the air compressor in the prior sentence. You may just want to say, "The water pump that keeps the water..."

Background Information
When the air compressor was first put in, the techs that installed it informed us that the room that we put it in was to small and not ventilated enough for this size of compressor. They said that we would need more ventilation to keep the heat out so that the compressor would operate better and not overheat. Previously we have put in two small exhaust fans in the wall of the room to blow the heat out of the room, but that not being successful enough for our needs. We know that during winter times the compressor is ok because of the cold weather but during the other half the year when it is hot, there is just too much heat in there.

> Comment [MSOffice34]: I would say "installed" again instead of put.

> Comment [MSOffice35]: Who is they?

> Comment [MSOffice36]: I'd remove this word. It puts it in present tense.

> Comment [MSOffice37]: Put a comma here

Effects of Problem
As you know, the compressor is a very important piece of equipment in this company. If we do not come up with a solution to the problem, it will affect every department in the company. Some departments can still run but not for too long. The paint department will be shut down completely since it is entirely run off of the air compressor. The other departments can run but will be slightly effected from not having their air lines for certain procedures. If we take care of this problem soon we shouldn't see a very big drop in production. But, if we don't fix the main problem and just fix the pump then the compressor will break down again and may require a very costly repair to get it running again.

> Comment [MSOffice38]: Insert comma

> Comment [MSOffice39]: In where?

> Comment [MSOffice40]: Insert comma

> Comment [MSOffice41]: It is not proper to start a sentence with a conjunction.

Causes of Problem
We know that the pump burned up prematurely, and just replacing it will not fix the problem permanently. Because the pump burned up prematurely there has to be a bigger reason that it broke. To find out some information on our specific compressor and what could have

> Comment [MSOffice42]: This sentence is weird to me.

caused this problem I have contacted the company we bought it from and talked to their technicians about what could have caused the malfunction.

We first looked at the room that it was installed in. The size of the room and what kind of ventilation it had compared to what the air compressor needed to run efficiently. It has come to my attention that the size of room does not really matter, but that the room has to stay cool enough to keep the compressor from overheating. Any temperature over 75 degrees is too much for the compressor and it will get hot and produce more moisture therefore over working the pump.

We then looked to see if the compressor was big enough for our purposes. Also, we needed to know what the settings were for the compressor. If it was storing enough air in its tank each time it refilled itself. If it is not, then I could be that we are over working the compressor by making it run more to try and keep up with the need for air in each department.

We also took a look at the maintenance schedule for our compressor. We need to know if the compressor was getting the oil changed and any filters changed on a regular basis. This could present a problem as well if the compressor is being neglected and not cared for properly. These four things are the most likely causes of our problem with our air compressor.

As a solution to this problem, we first should fix the water pump then take a look at the three ideas that I have found to be very good reasons why the pump failed prematurely and fix them accordingly. These repairs need to be done as soon as possible before any further problems with the compressor shuts down the plant until a costly repair is made. If you look at the end of the report you will see a chart showing which areas are impacted and how they are impacted.

Conclusion

I want to thank you again for taking the time to read my report about our current problems with the main air compressor. If we do not take this seriously and fix the problem in a timely manner we are looking at a very costly repair on our hands. I have concluded that there are four very likely causes to the water pump on the compressor to have burned out, and those are: the room that the compressor is in is not ventilated enough causing the compressor to overheat, the compressor is not big enough for our need, the compressor is overworking itself because the settings on it do not allow it to store up to its maximum air capacity, and the compressor is not getting the proper scheduled maintenance that it requires to run efficiently. The pump needs to be replaced and these other repairs need to be done before the compressor breaks down completely. If you need my help or have any questions about this report, feel free to ask.

Comment [MSOffice43]: Insert comma

Comment [MSOffice44]: You may want to start this sentence with "First, we looked at the room" that way you can order your troubleshooting instead of using "we" at the beginning of every paragraph.

Comment [MSOffice45]: You may want to make this two sentences. "...compressor. It will get hot..."

Comment [MSOffice46]: I think I understand what you're trying to say here, but the way its worded is confusing.

Comment [MSOffice47]: This is a run-on sentence.

Comment [MSOffice48]: I would delete this...."

Comment [MSOffice49]: "Following this report you will find a chart..."

Comment [MSOffice50]: Possibly say... "out. Those causes are:"

Comment [MSOffice51]: Maybe smooth this sentence out some... "It is my recommendation that the pump needs..."

Figure 2.5

Sample Student Peer Review (Jacob Horn, Peer Reviewer) *(continued)*

Chapter Summary

Understanding the different parts of the writing process and applying them to the writing challenges of the workplace will improve both your efficiency and your effectiveness as a communicator. Experience may offer shortcuts, but even the shortcuts are more workable when you understand the full spectrum of writing, revising, and producing. With practice, the discrete steps outlined in this chapter may blend into a more fluid process, but the principles remain reliable guideposts along the way.

End of Chapter Activities

ACTIVITY 1 Writing for Different Audiences

Directions. Please read the following scenario and complete the tasks in class. Type your responses and e-mail them to the instructor when you're finished. Be prepared to share in class.

Scenario. Last night something happened that caused you to miss your 9 A.M. interview appointment with a potential employer. The employer has narrowed the applicants down to you and another person; however, you've been led to believe you will get the job. This interview was to be your third and final one. The employer would make her decision after this interview. You REALLY want this job! Not only do you *really* want this job, but your parents have agreed to pay your deposit and first two months rent on a great apartment you've *already signed a lease for* because you were certain you'd get this job. Your parents live out of town.

Tasks. Pair with someone else in the class and work together to complete each task.

1. Write a letter to the potential employer explaining why you missed the interview appointment and asking for another chance to meet to explain why you're the person for the job. You *really* want this job.
2. Write a letter to your parents explaining why you missed the interview appointment and that you might not get the job after all. Remember, you've already signed the apartment lease and *need* their financial help.
3. Write a letter to your best friend explaining the situation.

ACTIVITY 2 How Writers Write

Find one or two people in the workplace (preferably someone in a job you'd like to have) and interview them on the role that writing plays in their job. Here are some questions to consider asking: How much writing do they do? When do they write? What percentage of their time do they spend writing? How much of their writing is informal? How much is formal? How often do they use e-mail to communicate? How much time do they spend speaking? What percentage of their total time do they spend communicating (writing and speaking)? How do they gather the information they use in their documents (research)? How do they document outside sources in their documents? In general, how well do they think their colleagues write? Speak? Write up your findings in memo format. Remember to single space your document.

ACTIVITY 3 Your Writing Process

Describe as completely as you can your own writing process. Do you have a favorite place to write? Do you start with the easy parts and move to the harder parts, or vice versa? Do you edit and write at the same time, or do you write all your ideas down before you start worrying about *how* to phrase a sentence or choose a particular word? What challenges you the most? The least? Add your own questions to this list. Be prepared to share with the class.

CHAPTER 3

Medium Choice

Overview

This chapter focuses on choosing an appropriate medium, whether oral or written, for your internal and external communication in the workplace.

Historically, the media for internal written communication have been the memo and e-mail; for external communication, the letter and report. Not so today. Technology has changed how we view communication and how we actually communicate. Now the two major medium categories are electronic and hard copy (or paper copy). Hard copy, for the most part, has taken a back seat to electronic mediums. In the electronic arena, Web conferencing, blogs, wikis, and podcasts are becoming commonplace, and in many organizations, e-mail and instant messaging have replaced the hard copy memo and letter for internal and external communication.

Many factors affect our choice of medium: speed of distribution, cost of distribution, length of time that the communication will exist, opportunity for quick feedback, and need for a legal signature, for example. Deciding whether to use an electronic medium or a hard-copy medium may come down to a privacy issue. As we know, electronic communication is not private! Employers have the legal right to monitor employee e-mail, both incoming and outgoing. Many court cases have been solved by tracing someone's e-mail messages. As a result, consider using a hard copy medium if communicating highly confidential information.

So, how do you decide which medium to use? Is it better, or just the custom, to communicate some topics via particular genres?

WHAT MEDIUM WOULD YOU USE FOR THE FOLLOWING?

You need to contact someone immediately.
Your reader needs a permanent record of the communication.
You want to establish a long-term business relationship with the reader.
Your reader is in another country.
Your message is complex and dense.
You want to make sure you communicate a friendly, collaborative tone.

As always, to make this decision, you must consider, among other things, the customs of the organization you're working in, the customs of the organization you're writing to, the purpose of the communication, the reader, and the ethics involved.

Electronic Mail, Memorandum, Letter

Whether writing an e-mail, a memo, or a letter in the workplace, certain standards and conventions apply. These standards and conventions vary in different cultures. Use the following writing strategies to guide you in writing just about any document. As you will see, no rule applies to every situation: As always, decisions flow from the context of the communication.

WRITING STRATEGIES

- State the main idea in the first or second sentence or at least in the first paragraph. Readers like to know what they're getting ready to read. *Caution: If the letter is giving bad news, you want to weigh carefully whether it's a good idea to begin the letter with the main idea!*
- Be brief, but not so brief that you're abrupt. In other words, give enough information so that your reader understands your message, but delete all words, phrases, and sentences that do not add to your meaning.
- Be accurate and factual. Avoid exaggerating or dramatizing your message or purposely misleading your reader—you will lose credibility quickly!
- Be reader friendly: Write to a person. Use, but don't overuse, the word *you* throughout the letter, particularly in the opening sentences. Phrase your message in terms of the reader's interests, needs, and expectations. Orient your reader.
  Exception: In letters of complaint or disagreement, the reader often perceives the "you" statements as hostile and accusatory. In these situations, use "I" statements instead.
- Use a natural, conversational style. Remember, you're writing to a human being. However, don't be too chatty or informal. Reading your letters aloud will help you spot any awkward areas. Asking a colleague to look it over is also a good idea.
- Be positive, pleasant, courteous, respectful. Avoid sarcasm, accusations, name-calling, blaming, or smart aleck remarks. You will only antagonize your reader with those comments. Treat your reader with the tone you'd prefer to hear from her.
- Be consistent in the tone you use. Shifting tone in your letter will only confuse the reader. Tone—the writer's attitude—is communicated through word choice, sentence length, abruptness, the amount of information given, and errors. The tone of a letter can be formal, casual, warm, cool, sad, hostile, friendly, patient, impatient, helpful, antagonistic.
- Use active voice instead of passive voice. Using active voice cuts down on the number of words and makes your writing stronger and more effective. *Remember, though, sometimes the passive is necessary and the better choice. In general, use strong verbs.*

- Use simple language. Choose the familiar word over the unfamiliar. *Remember, you're trying to communicate your message efficiently and effectively; using overly complicated words creates "fog" the reader has to work through.*
- Be specific in the words you use and in the requests you make. Don't refer to "the company's product" if you are writing about "how to put together the picnic table you sell."
- Avoid slang, jargon (often acceptable within organizations), clichés, exclusive language (elitist, sexist, racist, ageist), and stilted usage. Be extremely cautious about using jokes and humor. What you think is funny may not be funny to your reader.
- Use standard grammar, punctuation, and usage! Refer to a grammar handbook when necessary.

ELECTRONIC MAIL

E-mail has become, for most organizations, the standard way to communicate. It's easy, it's fast, it's convenient, and it's accessible to just about everyone.

E-mail is a hybrid, actually, a composite of the elements of a memo and a letter. Like the memo, it has a *To*, *From*, *Date*, and *Subject line*. Like the letter, it often has a *salutation* (Hi, Sam:), a *closing* (Sincerely), and a *signature block*. Also, much like the letterhead of the letter, the signature block includes all contact information.

An e-mail can be formal or informal. It can be short or long. The main message can be in an attachment or in the body of the e-mail itself. The genre of the message can be anything from an announcement of an upcoming meeting to a simple response to another e-mail to a memo to a letter to a multi-page report. See Figures 3.1 and 3.2 for sample e-mails written by former students Anthony Costa and Stephanie Johnson.

E-mail Writing Strategies

We sometimes become lax in writing e-mails because they are so quick and easy, but remember, your e-mail is *public*. It may get forwarded to any number of other people, both inside and outside the organization. So, whether the e-mail is a quick note to a colleague or a more formal report to a supervisor, include the following:

- A descriptive subject line
- A beginning, middle, and end
- Only information that relates to the subject line
- Complete sentences
- Standard capitalization, punctuation, spelling, and grammar
- An appropriate tone
- Your proofreading skills to make the message error-free

Database Software Troubles - Message

To... Chen, Jake

Cc...

Subject: Database Software Troubles

Dear Jake,

I'm having trouble with the piece of database software you gave me last week; it really isn't what I had in mind. I found that it doesn't have all the features I need to complete my job, and I wondered if you could suggest another that may be more helpful. If not, I have a few in mind, and I'd like to meet tomorrow to discuss them.

I need a piece of software that gives me a view of the database more like Microsoft Excel. The software you suggested will work, but the interface is a little too complex for what I need to do. This means that it takes me more time to enter data than it should really take. Also, the software you provided has a very cumbersome interface for making new tables. The software makes creating tables more time consuming than is really necessary, and I know there must be software that makes the ordeal simpler.

I have a few ideas for which software to use, but I'll need your permission to use them for the project. I've used them in the past and found them to be much more appropriate for this specific purpose. Let's set up a time to meet tomorrow to talk about which tools are a possible solution.

Anthony Costa
Undergraduate, Computer Science
Purdue School of Science
Phone: 317-371-8612

Figure 3.1

Sample Student E-mail

Here, Anthony is explaining a problem with existing database software. Note the first sentence identifies the problem, and the last sentence suggests next steps.

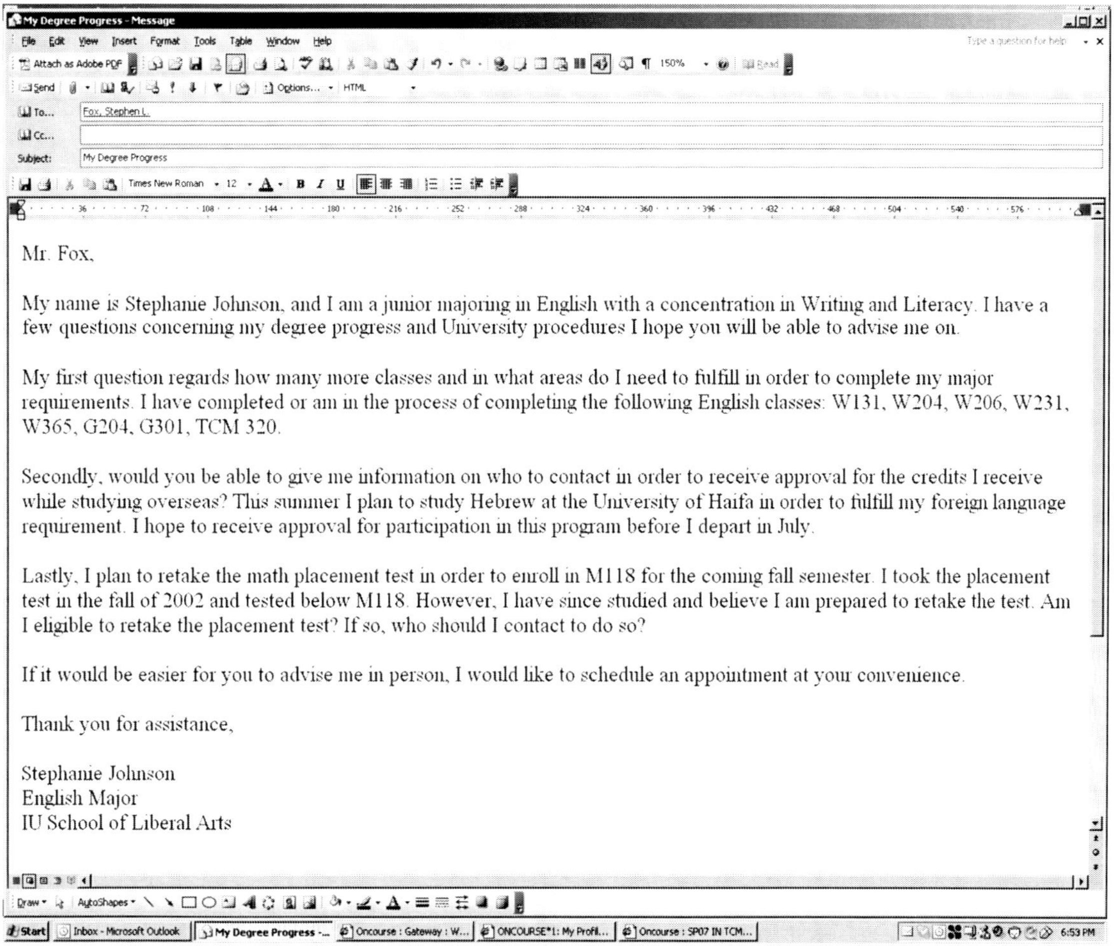

My Degree Progress - Message

To... Fox, Stephen L.

Cc...

Subject: My Degree Progress

Mr. Fox,

My name is Stephanie Johnson, and I am a junior majoring in English with a concentration in Writing and Literacy. I have a few questions concerning my degree progress and University procedures I hope you will be able to advise me on.

My first question regards how many more classes and in what areas do I need to fulfill in order to complete my major requirements. I have completed or am in the process of completing the following English classes: W131, W204, W206, W231, W365, G204, G301, TCM 320.

Secondly, would you be able to give me information on who to contact in order to receive approval for the credits I receive while studying overseas? This summer I plan to study Hebrew at the University of Haifa in order to fulfill my foreign language requirement. I hope to receive approval for participation in this program before I depart in July.

Lastly, I plan to retake the math placement test in order to enroll in M118 for the coming fall semester. I took the placement test in the fall of 2002 and tested below M118. However, I have since studied and believe I am prepared to retake the test. Am I eligible to retake the placement test? If so, who should I contact to do so?

If it would be easier for you to advise me in person, I would like to schedule an appointment at your convenience.

Thank you for assistance,

Stephanie Johnson
English Major
IU School of Liberal Arts

Figure 3.2

Sample Student E-mail

Here, Stephanie is requesting information about an academic program and her degree progress. Note the first two sentences orient the reader to the issues, and the last sentence ends on a positive, cooperative note.

Remember also to use these tools to make your message more accessible to the reader:

- Headings
- Bulleted lists
- Numbered lists
- Boldface
- Italics
- White space

A smiley emoticon is nice to use occasionally, but don't overuse it. Also, steer away from using e-mail shorthand (e.g., lol [laughing out loud]; btw [by the way]), since not everyone knows the language. You diminish the effectiveness of your message when your reader is concentrating on decoding the shorthand!

MEMORANDUM

As stated in the Overview, historically, the memorandum, whether printed or electronic, has been the informal, internal medium used in an office setting. Dictionary.com defines *memorandum* (often abbreviated "memo") as

1. A short note designating something to be remembered, esp. something to be done or acted upon in the future; reminder.
2. A record or written statement of something.
3. An informal message, especially one sent between two or more employees of the same company, concerning company business; *an interoffice memorandum.*[1]

However, memos are also commonly used to communicate informal messages externally. For example, a perfectly acceptable situation to use a memo is when communicating externally with committee members who may work in different organizations.

Here are the major parts of a memo:

To: Primary reader, Title (optional)
From: Writer, Title (optional)
Date: Date you're sending the memo
Subject: Exact topic you're writing about

The order of these parts varies; do follow the conventions of your organization. Some organizations will have a standard format that everyone is encouraged to follow; other organizations are more flexible. See Figure 3.3 for a sample memo by former student Andrew Hoblitzell.

Subject Line: An effective subject line conveys in a concise and specific way the information or action requested in the e-mail, memo, or letter.

Examples:

Poor: Meeting
Good: Benefits Meeting—March 5

Poor: Report Attached
Good: Smart Phone Recommendation—Blackberry

Memorandum

Figure 3.3
Sample Student Memo

To: Wanda Worley, instructor
From: Andrew Hoblitzell, student
Date: February 1, 2009
Subject: TCM 320: Course Reflective Memo

I have completed the final portfolio as you asked. This memo will discuss how I have accomplished the course goals and my own goals. The memo will also briefly discuss the design of my portfolio and my experience with the class.

Goal Completion
This table represents how the work over the entire course helped me achieve different goals.

GOAL	COMPLETION
1. Select, organize, and present technical information to readers in an organizational (technology, industry, or business) setting.	• Learned to use a variety of information in usable and persuasive ways. ◦ *Usability:* Presented instructions so that a novice could use them. Organized material with headings and other tools to help readers find things efficiently. Conducted usability testing to catch the biggest problems. ◦ *Persuasiveness:* Selected technical information and presented it in an accessible and persuasive manner.
2. Participate in peer review process with fellow students.	• Participated in peer review on multiple examples of technical communication. • Conducted usability testing to elicit "peer review" directly from the audience. • Collaborated with group members (somewhat like constant peer review): ◦ Participated in peer review with other groups. • Integrated review from mentors: ◦ Wanda ◦ TCM Writing Center

Figure 3.3
Sample Student Memo *(continued)*

GOAL	COMPLETION
3. Write using a reader-centered approach, which focuses on the needs of the reader.	• Considered the readers' characteristics and other factors constantly. ◦ Example: Conducted Usability Testing to elicit feedback directly from the audience.
4. Improve general writing and problem-solving skills.	• Learned new methods to use in presentation.
5. Create repeatable strategies or templates for business and technical communication.	• Wrote a variety of assignments with the same goals in mind: ◦ Stay reader-centered. ◦ Keep it simple, stupid! ◦ Integrate technical ideas into a communication for a novice. ◦ Review with peers.
6. Identify writing areas in which you can improve.	• Received a large amount of feedback with suggestions for improvement. • Produced my own substantial amount of feedback with suggestions for improvement.
7. Explore a wide variety of writing assignments.	• Worked with many different forms of technical communication: ◦ Personal Learning Statement ◦ Résumé ◦ Cover Letter ◦ Informative Abstract
8. Improve my overall writing skills.	• Satisfied with the results of the course.

Final Portfolio Design
I formed the final portfolio by considering the reader, the requirements, and the guidelines from class. I won't mention much else due to spatial constraints.

Class Experience
I had a wonderful time in this class. I know I still need to work on generating just what the reader wants and just in the time the reader needs it; but, I feel that my writing has improved significantly. I also wish that I had taken a version of the class that was a full semester.

Please contact me with any questions or comments.

Typically, the writer of the memo will sign or initial her/his name after the **From:** line. And, typically, memos do not have signature blocks, although the standard in some organizations is to have the writer of the memo sign at the end of the memo.

Although e-mail is used often in place of a hard copy memo, sometimes you may need a paper trail and so will want to use a hard copy memo. Perhaps, also, you're communicating confidential matter and want to use the hard copy memo for security or privacy reasons.

LETTER

Traditionally, the more formal letter has been the medium to use for external communication, that is, communication you're sending outside your own organization. However, letters are also often used internally to communicate very important information in a formal way. Traditionally, letter format and spacing are rigid. (*Note*: When using a template, you may need to manipulate some of the parts and spacing to meet the standard conventions.)

The **major parts** of a letter include the following:

- Company Letterhead—the company contact information: company name, address, phone number, fax number, e-mail address, Web site
- Date—the date you're sending the letter
- Inside Address—the reader's name, title, company name, and address
- Salutation—"Dear Mr. Smith:" (*Note*: A colon follows the name in a formal letter; in a personal letter, the comma is commonly used.)
- Body of Letter—the message, the content of your letter
- Closing—"Sincerely,"; "Cordially,"; "Best Regards," (*Note*: A comma follows the closing.)
- Signature Block—the writer's name typed, writer's title (*Note*: A comma separates the name from the title. Alternatively, the title may appear on the line underneath the typed name, in which case, the comma is not used.)

Typically, **spacing format** follows these specific conventions:

- Letterhead—number of lines depends on organization
- Date—two lines (or a double space) from where the letterhead ends
- Inside Address—number of lines from Date varies, depending on the length of the letter. If the letter is short, you'll want to add several blank lines; if the letter is long, you may need to double space only.
- Salutation—two lines (or a double space) from the last line of the Inside Address
- Body of Letter—two lines (or a double space) from the Salutation line
- Closing—two lines (or a double space) from the last line of the Body of Letter
- Signature Block—four lines (three blank lines) from the Closing line

See Figure 3.4 for a sample student job application letter.

Figure 3.4
Sample Student Job Application Letter

Notice how this job application letter explores the possibility of finding a job rather than applying directly for a specific job.

J. Student
12345 N. Pennsylvania St
Indianapolis, IN 46205
Cell Phone: 123-4567

Maureen Harrington, PhD
Department of Biochemistry and Molecular Biology
Indiana University School of Medicine
Indianapolis, IN 46202

Dear Dr. Harrington:

A few days ago a colleague directed me to a recent publication of yours. I read the online version of your paper in the journal *Luminescence*—the one about using luciferase to track tumor growth *in vivo*—and I was interested to learn more about the subject. As a researcher, my primary interest has been in performing fluorescence studies to better understand the metabolism of tumor cells, and I believe that your recent article will aid considerably in planning future fluorescence experiments.

For the past three years, I have been striving to learn as much about biology and chemistry as possible so that eventually I could contribute to the field with publications of my own, and I think that my attached resume demonstrates my dedication to the field. Ever since I began working as a student research assistant in a biochemistry lab, I have come to realize that experimental science is far different from anything conveyed in any textbook. Now that I am graduating, I am looking for a laboratory carrying out research in my field of interest, and I would like the chance to learn more about your work.

I see myself as articulate, intelligent, hard-working, and amiable. I have a natural curiosity about life and science, which gives me the motivation to grow and learn as a scientist. Consequently, I am seeking a position in academic science and I would enjoy the opportunity to hear about and to discuss your current fluorescence research.

Would it be possible to arrange a meeting at your laboratory or office? If so, please call me at (317) 123-4567 at your convenience, or e-mail me at jstudent@iupui.edu.

Sincerely,

J. Student

Reports: Formal and Informal

A common medium used in the workplace is the report. Reports can be classified into two large categories: formal and informal. Formal reports, often the result of projects that may span months or years, can be hundreds of pages. These reports generally follow a standard pattern:

- letter or memo of transmittal
- front matter (title page, abstract, table of contents, list of tables, list of figures)
- body (executive summary, introduction, discussion, conclusions, recommendations, references)
- back matter (appendices, bibliography, index)

Visuals (tables, figures, photographs, diagrams) are common and necessary in formal reports.

Informal reports, by contrast, may be a few paragraphs or a few pages. They can be in memo or letter format, printed or electronic. If electronic, the report can be the e-mail message or can be attached to a brief e-mail. Instead of a formal executive summary, an informal report has an introductory summary. An informal report has a body and a conclusion. Visuals are also used in informal reports, and if long or complicated may be included as appendices.

Using headings and bulleted and numbered lists is important in both formal and informal reports to make the material easily accessible to the reader. See Chapters 4 and 5 for a more detailed discussion on how to use reports to communicate your message.

Podcasting

Podcasting as a communication medium is growing in popularity. Because podcasts are simple to create and publish, people from all walks of life are using this medium to communicate: for example, your neighbor, doctors, lawyers, corporations, educators.

The word *podcast* derives from combining *iPod* and broad*cast*. Podcast is defined as "the Web-based broadcast of music which works with software that automatically detects new files and is accessed by subscription," according to *Webster's New Millennium*™ *Dictionary of English*.[2] Clearly, the definition has expanded to include more than just music. The act of podcasting, then, is sending a Web-based audio broadcast to subscribers over the Internet via a RSS (Really Simple Syndication) feed.

Podcasts are being used more and more in the workplace. For example, the CEO makes an important announcement about the future of the company, and the announcement is made available to all employees via a podcast. Hearing the announcement directly from the CEO ensures that everyone receives the same accurate information. To produce a podcast, you need a recording device (e.g., a computer or cell phone and a microphone) and a website to post your podcast. To listen to a podcast, you need a computer with Internet access and the ability to play MP3 audio files. Of course, if you don't want to be tied to your computer, you'll need an MP3 player for mobility to download as many podcasts as you want.

If you want to get started building your own library of podcasts on educational topics (many free), it's easy to find free downloads through iTunes by clicking on the "Podcasts" link under the "Source" or "iTunes Store" icon, depending on what window you have open. See Figure 3.5.

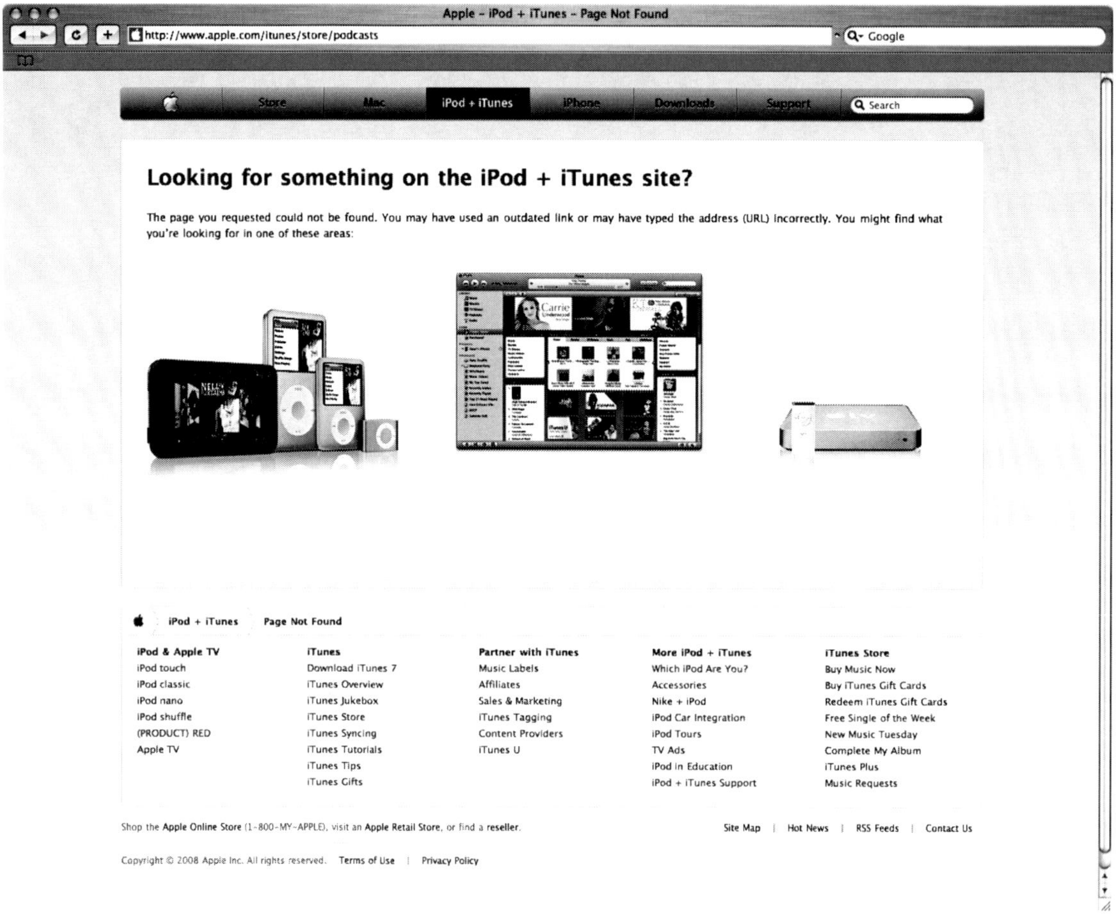

Figure 3.5
iTunes Podcasts Page

TRY THESE LINKS. SOME HAVE CATEGORIZED LISTS AND KEY WORD SEARCHING.

- http://www.apple.com/itunes/podcasts
- http://epnweb.org/
- http://podcastalley.com/
- http://www.podcastingnews.com/
- http://metrovoicemedia.com/index.html
- http://www.podcasting-station.com/

Audio digital recording has been around for a long time, but with the Internet and MP3 players, you can share your recording with the masses and contribute to the learning database on a large scale.

Wikis

A wiki is a medium used in many classrooms because students can use it to write collaborative documents, no matter the size. Professionals in the workplace are also finding the medium useful to write collaborative documents.

A wiki is a collaborative Web site where users can edit, delete, and add content, usually without restrictions.[3] The word *wiki* is an abbreviation of WikiWikiWeb, software developed by a computer programmer. *Wikiwiki* means quick in Hawaiian.[4] Wikis tend to work because so many people are involved that they self-regulate. The wiki software keeps a history of all changes made and even allows you to revert to a previous version if you don't approve of someone's edit.

The most well-known wiki in the United States is probably *Wikipedia,* a free online encyclopedia using wiki software. Wikipedia, started in 2001, currently contains 3,000,000+ articles in English, plus hundreds of thousands more in other languages.[5] Wikis are popping up on all kinds of subjects, such as recipes and the best vacation spots.

Blogs

Blogs come in all shapes and sizes. You can find political blogs, humor blogs, teen blogs, corporate blogs, celebrity blogs, software blogs. You name it and you can find it or create it!

A blog, short for (World Wide) Web + log,[6] allows the writer to publish a living personal or workplace journal. If you have something to say to the masses, a blog may be the ideal medium for you. Like a wiki, a blog is a collaborative Web site, but unlike the wiki, users do not edit, delete, or add content to what the author has written. Users can comment, however, on the author's reflections, information, experiences, or links. So the interaction and collaboration are there, which are two of the reasons blogging is a powerful communication medium.

According to Fernette and Brock Eide, blogs can:[7]

- Promote critical and analytical thinking
- Be a powerful promoter of creative, intuitive, and associational thinking
- Promote analogical thinking
- Be a powerful medium for increasing access and exposure to quality information
- Combine the best of solitary reflection and social interaction

In the workplace, a company blog may be a perfect medium for employees to express suggestions for improvements. With the ability to comment, a blog can also be a great collaboration tool for a department supervisor. Caution: Because a blog is public, we caution you to be very professional on any workplace blog.

Web and Video Conferencing

Web and video conferencing allows companies and organizations to hold meetings online. All that's needed is access to the Internet. Online conferencing is a common medium in the workplace. Web conferencing allows meetings to occur when travel is impractical because of time or money constraints.

Oral Presentations

Fear of public speaking—*glossophobia*—is one of the most prevalent and dreaded fears that exists. With the proper preparation and a lot of practice, though, you can reduce (maybe even overcome) these fears and improve your abilities. And, you will want to, because every oral presentation opportunity you have in the workplace is an opportunity for you to increase your worth in that organization. Your ability to orally communicate your ideas effectively is as important as your ability to write effectively. In fact, communicating orally in the workplace is ongoing: telephone conversations, informal meetings and discussions with colleagues and supervisors inside and outside the organization, discussion at staff meetings, formal meetings and formal presentations inside and outside the organization. The higher up you are in the management rank, the more you will use your public speaking skills.

So, when you are called on to make a more formal oral presentation, whether in the classroom or the workplace, show enthusiasm and confidence, repeat key insights, and keep it simple by focusing on a few key ideas that you want the audience to remember from your presentation. Giving an oral presentation involves many of the same steps as writing a technical report: (1) planning,

(2) researching, and (3) organizing the content. Beyond these similar steps, you will need to (1) prepare the visuals for the presentation, (2) practice, and (3) deliver the presentation.

PLANNING

You've determined your topic and purpose, and now it's time to figure out who your audience is. What does your audience already know about your topic? What does your audience want to know about your topic? Think of your audience as having different layers of knowledge about your topic. Some of them may be experts; some may know something about the topic; some may know little or nothing about the topic. Decide which layer you will aim your presentation, but make sure you give everyone something. You have to decide what you want them to know.

Planning also involves choosing the most effective presentation technology. If you're presenting to a large group, a digital projector is probably the best alternative. However, if you're presenting to a department staff, the white board or a flip chart may better serve you. The overhead projector, although old-fashioned, is still used occasionally. And if it's an academic presentation, you may want to use a poster to display your information.

Will you give the audience a handout of your information? If so, you need to plan the handout. And finally, if you're presenting with others, you will want to plan the logistics of the presentation; especially important is deciding how you will transition between speakers.

RESEARCHING AND ORGANIZING CONTENT

Just as you have to gather information for a technical report you're writing, you must gather information for your oral presentation. When you have your material gathered, you need to organize it in the way that best meets your audience's needs and your purpose. Similar to a written technical report, the presentation should have an introduction, body, and conclusion. In the introduction, forecast what you're going to tell them; in the body, tell them what you want them to remember; in the conclusion, summarize what you told them. So the adage "Tell them what you are going to tell them, tell them, and then tell them what you told them" holds true for an oral presentation as well as a written technical report.

PREPARING THE VISUAL AIDS

You must next decide on and prepare visual aids you plan to use in your oral presentation. Visual aids can refer either to the medium you choose to display your message (such as PowerPoint, white board, poster) or to any illustrative material you use via that medium to get your message across (for example, photos, graphs, tables, screen shots). Visual aids support your message and help focus your listeners' attention. Refer to Chapter 7 for more information on illustrative material whether used in a technical report or an oral presentation.

PRACTICING AND DELIVERING YOUR ORAL PRESENTATION

Practice, practice, practice. Practice in front of a mirror, practice in front of friends, practice using the technology, but practice. Delivering the presentation is, for most of us, the challenge. These tips may help:

- Be prepared.
- Let the listeners know upfront what you're getting ready to tell them.
- Make sure the organization of your key ideas is clear to the listeners.
- Use transitions often throughout the presentation; listeners need to know how one idea relates to another.
- Visualize yourself giving a successful presentation.
- Use body language effectively:
    - Dress appropriately.
    - Stand up straight.
    - Move and gesture naturally; don't fidget.
    - Use open hand and arm gestures when possible.
    - Make eye contact.
- Keep voice, rhythm, and tone appropriate to audience and purpose:
    - Articulate clearly; don't mumble.
    - Slow down.
    - Pause to eliminate *um*, *uh*, and *ah*.
- Use notes effectively:
    - Let slides remind you of your topics and subtopics.
    - Don't read notes or slides word for word.
    - Number note cards if you use them; if you drop them, you can put them back in order more easily.
    - Use large font size and few words.
    - Don't panic if something breaks.
    - Plan to talk, on average, two minutes per slide.
- Involve the listeners, when possible.
- Manage the question and answer period (Q and A):
    - Anticipate potential questions.
    - Be brief.
    - Stay focused.
- Following the Q and A, make a final comment.

Remember, though, talking to an audience is different from writing to an audience. Different conventions drive the delivery. The most common medium used to deliver an oral presentation is PowerPoint, often accompanied with a handout. Others, however, are effective: overhead transparencies, white boards, flip charts.

PowerPoint

As a student and employee, you've probably seen your share of good and bad PowerPoint presentations. However, the tool is an effective medium if used properly. In the workplace, PowerPoint is often used in meetings when the topic to be discussed is somewhat complicated or dense or when the audience is a large group of people. A PowerPoint can be especially effective if you're sharing a new plan or procedure in a meeting and need to focus the participants. Many people need a visual to look at while listening. The power of PowerPoint is that it doesn't have to be boring or linear; with a dynamic PowerPoint, your topic comes alive.

Possibly the biggest mistake people make with a PowerPoint presentation is putting too much information on the PowerPoint slides. The PowerPoint is not your presentation; it is a supplement to your presentation. If everything you have to say is written on the PowerPoint slides, you'll have many people leaving the meeting grumbling under their breath, "Why didn't he just give us the PowerPoint to read?"

Here are a few tips to remember as you prepare your PowerPoint slides:

- Limit the amount of text you put on each slide.
- In general, use words and phrases, not sentences. Sentences tend to make us want to read what's there instead of talking with our audience.
- Whenever possible, use illustrations instead of words. A picture is worth a thousand words! A bar chart may be just what you need to illustrate your point.
- In general, plan a slide for each two minutes of presentation. So if you're presenting for ten minutes, you'll want five slides.
- Make visuals readable, clear, and large enough for the listeners to see.
- Make visuals simple. Your purpose is to communicate.
- Make sure the slide title fits the content of that slide.
- Use a readable font style and size and be consistent throughout the presentation. Many experts suggest using a sans serif font (Arial, Tahoma, for example) for slide titles and a serif font (for example, Times New Roman, Courier) for slide text.
- Use parallel structure on each slide.
- Keep the background simple. Your message should be the focus, not the slide background.
- Use color with care. If your audience is international, remember that color means different things in different cultures.
- Make the PowerPoint error-free!

Whether you use text or visuals to depict your message, let your audience and purpose guide your PowerPoint design.

Face-to-Face Conversation

Of course, face-to-face conversation is the most effective medium to use in certain circumstances. For example, if the topic is highly sensitive, you probably want to meet with the person. Or, if you've never met the person and you want to establish a long-term business relationship, a face-to-face meeting will work better than a letter, for example. Another situation in which a face-to-face meeting works well is when a problem needs to be solved. Meetings are common in the workplace (some would say too common) and provide an opportunity for community building, information sharing, and decision making. Let the situation, purpose, and audience determine whether it's necessary to meet the person face-to-face.

Chapter Summary

This chapter has emphasized the variety of options that you have at your disposal when you need to communicate. Paper, electronic, and multimedia all have their functions in the workplace. Your job is to understand both the advantages and limitations of your options, and then to exercise them intelligently and strategically. Whether you choose to send a formal letter or initiate a blog, you will be demonstrating your mastery of that medium.

Sources

1. memorandum. (n.d.). *Dictionary.com Unabridged (v 1.1)*. Retrieved March 14, 2007, from Dictionary.com, http://dictionary.reference.com/browse/memorandum
2. podcasting. (n.d.). *Webster's New Millennium™ Dictionary of English, Preview Edition (v 0.9.6)*. Retrieved March 14, 2007, from Dictionary.com, http://dictionary.reference.com/browse/podcast
3. wiki. (n.d.). *The American Heritage® Dictionary of the English Language, Fourth Edition*. Retrieved March 14, 2007, from Dictionary.com, http://dictionary.reference.com/browse/wiki
4. wiki. (n.d.). *The American Heritage® Dictionary of the English Language, Fourth Edition*. Retrieved March 14, 2007, from Dictionary.com, http://dictionary.reference.com/browse/wiki
5. wikipedia.org. Retrieved January 24, 2010, from wikipedia.org, http://en.wikipedia.org/wiki/Main Page
6. blog. (n.d.). *Online Etymology Dictionary*. Retrieved March 14, 2008, from Dictionary.com, http://dictionary.reference.com/browse/blog
7. Richardson, W. (2006). *Blogs, Wikis, Podcasts, and other powerful Web tools for classrooms*. Thousand Oaks, CA: Corwin Press.

End of Chapter Activities

ACTIVITY 1 Memo Practice Writing

Organize the following information into a logical sequence and then create a memo with appropriate form and headings. Address memo to the entire staff of XYZ corporation:

- XYZ is in the process of instituting a systemwide quality control program.
- You're the chair of a committee—one of several committees set up to determine ways XYZ can accomplish this goal.
- Your committee needs to know exactly how people spend their time while on the job. Your committee has designed a form for employees to use to record their activities.
- Employees are to complete the log, accounting for every minute of their day every day—for a four-week period.
- All employees have been told about the quality control program. They do not know about the time log.
- A side benefit of keeping the log is, of course, that individuals can see how they use their time and will, as a result, be able to make changes that affect quality control for the entire system.
- Everyone must keep a log.
- Employees have been told that you, as the chair of your group, will be asking them for information and that they are to cooperate fully.
- Employees are to give completed logs to their supervisors at the end of each day.
- Supervisors are to hold the logs until the end of the week and then give them to you. Employees can make copies for their own use.
- Your committee will analyze the data and let everyone know the results.
- All employees were given the opportunity to serve on a committee.

ACTIVITY 2 Wikis

Complete in the computer lab. In small groups of three to five people, explore *Wikipedia* (www.wikipedia.com). Each person in the group should research a different topic. What are your thoughts about the credibility of the information you found in *Wikipedia*? Discuss as a group. Query a librarian to see if the librarian views *Wikipedia* as a credible source. Check any other sources you can find on the credibility of *Wikipedia*. Be prepared to present your findings to the class.

ACTIVITY 3 Test Your EQ

Make a list of any problems you find as you read the following e-mail message. Be prepared to discuss in class.

Subject: FYI
Date: 08-05-25 11:58 EST
From: Lockwood Management
To: Office Managers, Barlow Building
Attached file:

On Tuesday, May 23rd, computer equipment was stolen from Global Mailing, a tenant on the 7th floor. (Global Mailing is the company that gives those legendary Christmas parties with the unforgettable rum punch!) Our security camera recorded a man leaving the building. Apparently the thief entered the unlocked suite, placed the computer in a moving crate and loaded it onto a van parked at the rear loading dock. The Boston Police were contacted and a report was filed. The police informed us that there is a man posing as a professional mover who enters offices, steels computers and other office equipment as well as wallets. The tape with the photo of the suspect is available at the guard's desk. Please stop by and take a look. If you recognize this man, have any information regarding this theft, or if you experience a theft in your suite, please contact us, and the Boston Police. Please remember never to leave your reception area unattended and to screen visitors carefully. Remember to keep your suite doors and desks locked. The attached file is our 30 page manual containing information on security procedures. Please familiarize yourself with the relevant material. BTW, the electricity will be shut off at noon today (May 25) and remain off for two hours for maintenance. If using a computer, take appropriate steps so you won't lose data during this power outage.

ACTIVITY 4 PowerPoint Practice

In small groups of three to five people, choose a chapter in this textbook and prepare a PowerPoint presentation for the class on the content of the chapter. Be prepared to discuss the decisions you made after your presentation.

CHAPTER 4

Organizational Strategies

Overview

So far we have talked a lot about the elements of effective informal reports and the various decisions a person makes when writing them. This chapter will concentrate on practical advice and techniques to use in producing an informal report.

The process does not have to be complicated or intimidating. The operating principle of good technical workplace writing is to keep the writing simple, concise, and clear. Like many other activities, good preparation is key to accomplishing that goal.

We begin with some general advice. The first is commonsensical. An effective informal report—like most good writing—has a beginning, a middle, and an end, each with its own function:

- *The beginning establishes the reason for the communication, its circumstances and purpose, and an overview of what to expect.* Generally, it consists of one or two paragraphs.
- *The middle contains the meat of the communication.* The main points are not only introduced, but explained here—this section contains the details. By its nature, this part of the report is the longest and may provide the largest challenge. Your job is to explain thoroughly the points you want to make, keeping in mind your reader's needs and patience level. You want to explain, not overexplain nor underexplain. You want to be clear and organized. You want your points to flow together.
- *The end brings the report to a logical close.* Because by its nature an informal report is usually a short document, you may not have to repeat each idea that the report has introduced, but you will want to signal to your reader that the report has ended. Often writers use the end of the report to express their willingness to continue their involvement, to answer any questions, or to take the next steps. Like the beginning, the end is often one or two paragraphs.
- *Attachments (as needed) add explanatory or supporting information.* They are referred to in the report.

Beginning

The beginning establishes the **context** for communication. Remember, people are busy and have many activities going on simultaneously. Begin your memo with an **orienting sentence** that will establish the context of your communication. *(You asked me to. . . . Before you meet with the customer, you should keep these ideas in mind. . . . As we discussed in our meeting yesterday. . . .)*

The beginning also establishes the **purpose** of communication. Don't keep your reader guessing about why you have sent the memo. Use the appropriate **verb** that will guide the discussion. If you are going to compare the offerings of three vendors, use the word *compare;* if you are intending to analyze the factors contributing to a problem, make sure the word *analyze* appears in your first paragraph. If you are recommending, indicate that you intend to recommend. *(I have compared the three service options on the basis of distance from the office, time and materials, hourly rate, and night and weekend availability. My team has analyzed the data to determine the reasons for the drop in revenue. Keeping these factors in mind, I recommend that we realign the service department with operations.)*

Finally, the beginning sets the reader's **expectations** for the number of points to follow as well as for the importance of those points. Make it easy for the reader to anticipate. *(I will first outline the three factors that make this issue important and then suggest a possible solution. . . .)*

Middle

The middle establishes the **order** of points. Make the organizing principle clear from the start. Use transitional words that will help the reader keep on track. Have a strategy for the order in which you present your ideas. Should you start with the most important point first, on the premise that your reader may never read the whole report, or should you start with the easy idea and build to the more complex? As you make those decisions, keep in mind the nature or purpose of the communication, the available information, the reaction you expect from your reader, and your knowledge and history of the audience as reader. *(The first, and probably simplest, reason. . . . In addition to the extra expense, another concern. . . .)*

The middle also provides the reader with enough **data** and **examples** both to explain the reasoning and to support the points. Again, transitional elements help the reader follow your train of thought and stay focused. *(For example, the attendance figures for the month of January. . . . Unlike the Austin office, the Cleveland office. . . . The cost of utilities, on the other hand, . . .)*

End

The end brings the report to a logical close. It may summarize the findings, may simply articulate the conclusion to which the report has been building, or may offer the next steps to be taken. *(Having considered all the factors, I recommend that we initiate flexible hours on a limited, trial basis as soon as we can work out the details.)*

The end also reiterates the **good will** of the writer, expressing offers of further help or clarification or of willingness to pursue a course of action. *(If you have any questions or concerns, I am available any time. Please call me if you need anything else. I'll be happy to sit in on the call with you if you prefer. . . .)*

Executive Summary and Introduction

Often at the beginning of a long report, the first section will be labeled the *Executive Summary*. The name explains its purpose: It is a summary of the main points of the report intended for executives who probably do not have the time or inclination to read the whole report in detail. Because the executive's role often centers on the financial and management pieces of the business, the executive does not necessarily have to understand every tiny detail that the implementation specialist or the technician needs to know. So the Executive Summary, a smaller, capsule version of the report, contains the major purpose, conclusions, recommendations, and financial details.

Note that this convention is primarily used to summarize long, formal reports. The *Executive Summary*, however, does not replace the *Introduction* of the report. In an informal workplace report of only two or three pages, a formal executive summary may not be necessary. An *Introduction*, stating the context, purpose, and intention, often suffices. Some companies, however, may require an Executive Summary for every office communication, so you will want to be familiar and comfortable with the term.

Requests for Proposals (RFPs)—formalized documents sent to vendors that specify what a company is looking for in making a major purchase—often require an Executive Summary. White papers (defined as *authoritative reports*) contain them as well.

You will find an example of a student report containing an Executive Summary at the end of this chapter (Figure 4.1).

Preparation

The second piece of advice is to prepare well. If you know in advance both what you want to say and how you want to say it, the actual writing task will seem less daunting. Using this idea as our guiding principle, we have created worksheets that will help organize your thoughts and keep you focused on the challenge at hand.

We believe that getting your ideas on paper (or screen) makes them more real and relevant—also workable. The worksheets found in Appendix A will help you describe who your audience is, what circumstances have instigated the report, and what organization will best serve your purpose. This preparation will expedite your actual writing time and help make your report clear and on target. Each type of report has three corresponding worksheets: Communication Strategy Worksheet, Design Worksheet, and Peer Review Worksheet.

COMMUNICATION STRATEGY WORKSHEET

The first worksheet is the Communication Strategy Worksheet; its purpose is to help clarify your knowledge of the situation driving the report and of the audience receiving it. The worksheet questions will help you identify the elements of the situation that should shape the information you need to present. We have spent some time talking about the abstract notions of audience and purpose; now on the worksheet, you have the opportunity to translate those abstractions into concrete realities. Who, what, where, when, why, and how are all the underlying questions you must answer.

DESIGN WORKSHEET

The second worksheet is the Design Worksheet. Having gathered your information on the Communication Strategy Worksheet, you now need to think about the organization and shape of your report. The diamond shape reflects the natural shape of most reports. The beginning and ending paragraphs certainly are important, but the guts of the report are in the middle, where the ideas must be developed, explained, and perhaps even defended.

Use the Design Worksheet to make sure that you have covered the important aspects of the situation. Of course, every report may not contain every possible approach covered on the Design Worksheet, but the worksheet offers some ideas and strategies for organizing and composing your report. Think not only of the content of each section, but also the proportions of each. Once you have completed the Design Worksheet satisfactorily, you will be ready to write your first draft. In fact, you will probably find that the draft is half written!

PEER REVIEW WORKSHEET

The third worksheet is designed to help guide the peer review. Specific questions address the concepts—both general and detailed—discussed in this chapter. Although the primary purpose is to help guide the reviewer, you certainly can use it to self-check, thus becoming your own first reviewer. After you complete your first draft and take a break from your report, come back and fill out a Peer Review Worksheet. How does your report measure up? As you will see in the samples in Appendix A, the worksheet may help you identify areas of your report on which you would like specific feedback. The worksheet questions will help you pinpoint and articulate concerns in the report text itself. See Chapter 2 for samples of peer reviews.

Chunks and Headings

Completing the Communication Strategy and Design Worksheets will provide information and concrete data with which to work. You have passed out of the purely speculative stage of your report and into the actual production stage. Now the task is to organize the information, and again, that job does not have to be overwhelming; just follow the logical path.

CHUNKS

First, write out your main idea, the purpose behind the informal report. That main idea will probably contain the guiding verb of the report, such as *compare* or *analyze* or *recommend*. Next, identify the details that will explain or support your point. Think of each of these ideas as a chunk of information—a block of words that offers the explanation and details of your research.

Look at your Design Worksheet, take each chunk, and write out the points you want to make. Decide your order of presentation. You may take a chronological approach, from what happened first to what happened next. You may decide to begin with your weaker points so that you can build to your stronger ones. You may choose the opposite route and devote the beginning of your argument to the strongest point and move on to the least important. If you are recommending, you may want to offer first the least satisfactory alternatives to build your case to the most viable one, or you may start with your first choice to emphasize its importance. Choose and draft.

HEADINGS, SUBHEADINGS, AND SUB-SUBHEADINGS

Now your report is finding its feet—you should have a sense of the beginning, middle, and end. Notice that these natural divisions lend themselves to being divided by headings. Though you will not necessarily label your headings *Beginning, Middle*, and *End*, those words do describe their basic functions. You may begin with *Introduction* or *Background Issues*, or some similar title; then lead to *Contributing Factors* or *Comparative Studies;* and finish with *Final Recommendation* or *Summary of Findings*. Whatever accurate label you give, the function is clear.

Notice, too, that subheadings, particularly in the middle section of the paper, may serve as obvious markers to help the reader follow the path of the report. Subheadings act as road signs, signaling the beginning of a new point. Used effectively, they help the reader quickly identify the main points and quickly absorb the most important information. They offer a psychological as well as a physical break in the text so the reader feels less intimidated by a lot of information. Used in conjunction with precise transitional expressions, they help make your report readable and user-friendly.

As you can see, the strategic use of formatting is important in technical workplace communication. When you consider how much written information people encounter in the workplace every day, anything you can do to make your reader's job easier will be appreciated. Using good headings and subheadings is a practical and relatively easy technique. Looking at your chunks of information, decide first what major headings you want to use, and then apply useful subheadings (if they would help) to your developmental ideas.

If you choose to use a ready-made template or style sheet, be aware that some of them are poorly designed and may be difficult to work with.

Heading Labels and Variations

Deciding what headings and subheadings to use is a challenge, but, as we have advised before, use common sense and sound judgment. You might pick generic headings or customize them to specifically identify the points in the report. We recommend the latter.

In Table 4.1, you will find some suggested titles, which lend themselves either to serve as the heading or to be customized for your report. Use this list—mix and match—as a starting point for your road signs.

Table 4.1 Sample Headings

Beginning	Middle	End
Introduction	Discussion	Conclusion
Issue	Contributing Factors	Resolution
Problem	Elements	Solution
Background	Alternatives	Recommendation
History	Factors	Summary
Introductory Summary	Criteria	Suggestions
Challenge	Analysis	Action
Situation	Comparison	Proposal
Review	Causes/Effects	Next Steps

Table 4.2 shows examples of how to move the heading from its most generic form to *a more specific label* to a *very specific, descriptive label*.

Table 4.2 General to Specific Headings

Beginning	Middle	End
Introduction	Discussion	Conclusion
Current Travel Policy	*Significant Concerns*	*Implications for Company*
Travel Policy Needs Revision	**Four Factors Contribute to Issue**	**Company Should Implement Changes**
Problem	Elements	Solution
Absenteeism	*Data and Trends*	*New Policy*
Company Is Experiencing Increased Absenteeism	**Data Indicate Problem Is Worsening**	**We Need to Change the Policy**
Introductory Summary	Criteria	Recommendation
Cell Phone Selection	*Critical Criteria*	*Recommended Vendor*
Recommendation for Change in Cell Phone Vendor	**Four Critical Criteria**	**Recommendation: Blum Wireless**

Avoid *lone* headings or subheadings. That is, if you have one heading, make sure you have a second one. The same goes for subheadings. Also avoid ending the page with a heading, called a *widowed* heading. The general rule of thumb is to include at least two lines of body text under the heading. If you can't, move the heading and body text to the next page.

Finally, we offer this advice about designing your headings and subheadings: Stay consistent. Headings and subheadings (and sub-subheadings) represent a hierarchy of relationships. Headings obviously are the most important, so they should all match in type face, font size, position on page, and any defining formatting like bold or italics. Subheadings support the headings, so they look different from the headings, but identical to each other. By looking at the design of your document, by the look and placement of your headings, your reader should be able to determine your most important points and your secondary, supporting points.

Bullets, Lists, White Space

In keeping with the idea of making the report easy to read and follow, you can also break up big chunks of text by using bulleted or numbered lists. Bulleted lists are generally used when all the items have approximately the same importance and effect; numbered lists—by the very nature of assigning numbers—usually carry some implied order to the importance. (Think of Dave Letterman's "Top Ten Lists"; we expect the items in the list to get funnier and funnier as the numbers decrease to number 1.)

Here is an example of how a famous piece of writing—the Preamble to the Constitution of the United States—appears in its original paragraphed (chunked) form and then how a modern writer might create it in a bulleted list:

ORIGINAL

"We the People of the United States, in Order to form a more perfect Union, establish Justice, insure domestic Tranquility, provide for the common defense, promote the general Welfare, and secure the Blessings of Liberty to ourselves and our Posterity, do ordain and establish this Constitution for the United States of America."

COMPARE TO

"We the People of the United States, in Order to

- form a more perfect Union,
- establish Justice,
- insure domestic Tranquility,
- provide for the common defense,
- promote the general Welfare,
- and secure the Blessings of Liberty to ourselves and our Posterity

do ordain and establish this Constitution for the United States of America."

Notice how separating the several goals of the founding fathers into bullets makes easier reading. The text is less dense because this version uses more white space. Notice, also, that the reasons all begin with the infinitive verb form (*parallel structure*, see Chapter 8), which also emphasizes the equality of the ideas.

Like anything else, numbered and bulleted lists can be overdone, but keep the technique in mind when you sense that the chunks of text look too blocky.

Sample Student Reports

The following pages contain a report with an Executive Summary written by former student Andrew Hoblitzell (Figure 4.1). Following Andrew's report is an informal report written by former student Ken Jenkins (Figure 4.2). Notice how both reports use the techniques just discussed to produce effective, accessible informal reports.

Figure 4.1
Sample Student Report

ABC Corporation

Memo

To: Bob Smith, Financial Manager
From: Andrew Hoblitzell, Computer System Administrator
Date: 6/14/2007
Re: Proposal for an Adaptable Computer Security Budget

Executive Summary

ABC Corp. has consistently had trouble recognizing growing computer security threats and the need for an adaptable computer security budget. According to CERT (2007), a federally funded Internet security research and development center based at Carnegie Mellon University, 137,529 computer security intrusions were reported to them in 2003 (p. 1); and, this number continues to grow rapidly every year. After an initially low analysis cost, ABC Corp. will be able to optimize its spending for security according to a new methodology and easily gain a measurable return on analysis costs. I believe it is very important for ABC Corp. to seize this opportunity.

Problem

The first step for organizations with a computer security problem is recognizing that they have a problem. Recognizing computer security problems becomes increasingly important as vulnerabilities proliferate. CERT (2007), a federally funded Internet security research and development center based at Carnegie Mellon University, captured the growing computer security problem in statistics released in 2007 (see Figure 1).

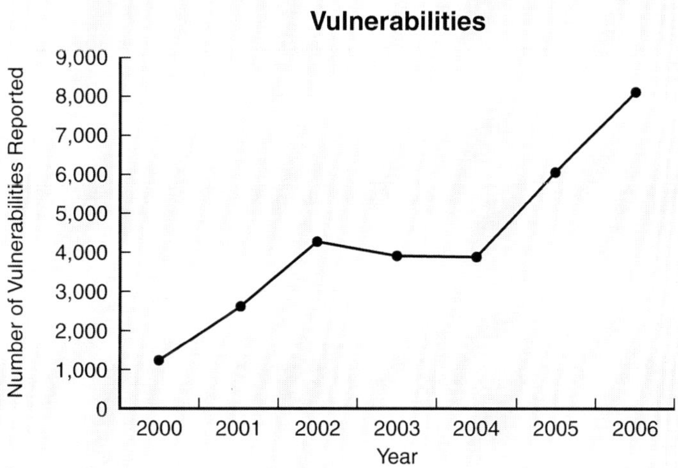

Figure 1

Figure 4.1
Sample Student Report *(continued)*

For a table and more information about the growing number of computer security vulnerabilities, see Appendix A. For information about the growing number of actual intrusions, see Appendix B.

According to Jaikumar Vijayan (2003), a senior editor at *Computer World* magazine, the largest obstacle to implementing computer security at most companies is inadequate funding (p. 1).

Once we have recognized that there is a problem with our current computer security strategy, the problem becomes developing a budget which minimizes total cost for the company. This issue is discussed next.

Objectives

Developing an adaptable computer security budget is important for the bottom line of ABC Corp. because we are currently spending our funds inefficiently. Stuart E. Schechter (2005), a member of the Information Assurance group at MIT Lincoln Laboratory, advises developing "a process that reduces the number of potential adversaries" (p. 1).

For ABC Corp., this means developing a computer security strategy that reduces the number of enemies from the start. To accomplish this, we need to:

- Realize that there is a computer budget problem at ABC.
- Determine a new computer security budgeting methodology for ABC.
- Implement the new computer security budget after further discussion.

Solution

I propose to complete these objectives by establishing a new methodology for formulating ABC Corporation's computer security budget. To do this, I suggest the following:

- Conduct research about the failure rates and observed damages for each of our security layers.
- Determine total costs for all of our current security layers.
- Calculate an optimal RROI (Risk-based Return on Investment) based on risk reduction rather than increased productivity.
- Apply advice listed in a paper from the field of econometrics:
  (a) Stay up to date on patches.
  (b) Develop a simple alarm system that notifies the vandals we are aware of them.

After carrying out these steps, ABC Corp. will be able to determine quickly how much money it has saved by calculating new observed damages and bypass rates and comparing them to the old ones. Under the methodology proposed, ABC Corp. will be able to guarantee that it is making a profitable investment until the RROI ratio falls below 1. This topic will be explored more in a later section.

Figure 4.1
Sample Student Report *(continued)*

Procedures

I propose to establish a new computer security budgeting methodology based on research applied at Lawrence Berkeley National Laboratory (steps 1–3) and econometric research from Stuart E. Schechter, a member of the Information Assurance group at MIT Lincoln Laboratory. I propose the following steps:

Step 1: Gather the necessary data for the analysis.
Step 2: Estimate bypass rates and perform calculations.
Step 3: Evaluate the findings.
Step 4: Synthesize the findings with insight from econometrics.

For a more detailed layout of steps 1–3 (Arora et al., 2004, p. 2), see Figure 2.

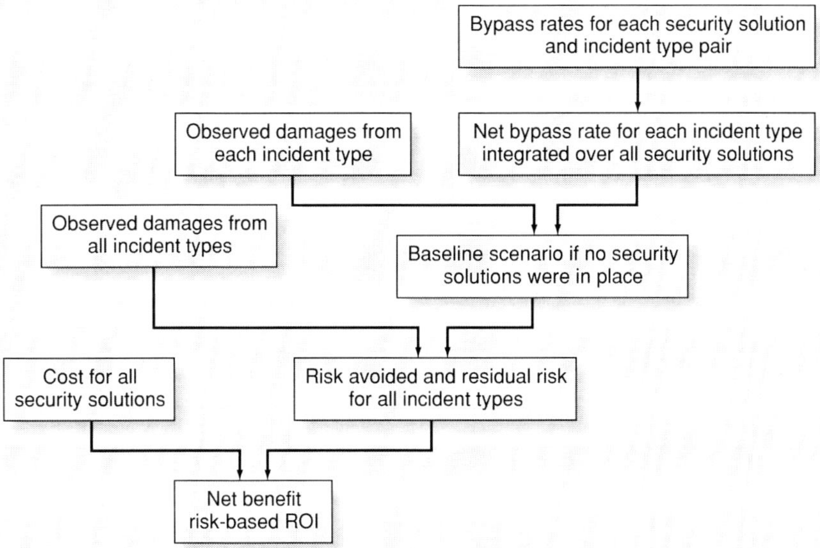

Figure 2
Basic Process for Estimating Risk-based Return on Investment
(Source: Arora et al., 2004, p. 2)

Step 1: Gather the necessary data for the analysis. The first data we need to collect is observed damage, or the amount of money ABC Corp. has lost due to security breaches in a given period of time. Next, we will need to determine the cost and bypass rate, a measure of the effectiveness of our security layers, for each different security solution.

Figure 4.1
Sample Student Report *(continued)*

Step 2: Estimate bypass rates and perform calculations. After someone familiar with the security situation estimates the bypass rates for each layer of security, the net bypass rate will be calculated by simply multiplying all of the individual bypass rates together.

We will then calculate incident risk, or the amount of money each security type would cost ABC Corp. if no security solution was in place. We calculate the incident risk for each layer by dividing observed damage for that incident by the net bypass rate for that incident. This figure represents the amount of money ABC Corp. would be losing if no security solution were in place for this incident type.

We will then calculate the baseline scenario, or the amount of money ABC Corp. would pay for all security intrusions, by simply adding together all of the incident risks.

Step 3: Evaluate the findings. Next we will calculate the RROI, or risk-based return on investment, by subtracting the sum of the residual risk and cost from the baseline scenario and then dividing by the cost. Writing this as an expression yields Figure 3:

$$\text{RROI} = \frac{\text{Baseline scenario} - \text{Residual risk} - \text{Cost}}{\text{Cost}}$$

Figure 3
Calculating Your Organization's RROI
(Source: Arora et al., 2004, p. 3)

If the calculated RROI is greater than one, ABC Corp. avoids more risk by investing in this security solution (in other words, we should definitely invest our money in this particular solution). If RROI is between zero and one, ABC Corp. should continue to invest until we find a diminishing return that is acceptable to us.

Step 4: Synthesize the findings with insight from econometrics. Spending for each of the security layers should also take into account the results of Schechter's research (2005), that an organization needs "a process that reduces the number of potential adversaries, reduces the value that can be obtained through successful attacks, increases the risks incurred by adversaries during attack, and raises other attack costs" (p. 1).

For example, ABC Corp. could implement a simple alarm system that would notify the vandals that we had found them on our system and that we were automatically referring their intrusion to the proper authorities. We could then automatically refer their intrusion to the proper authorities.

Figure 4.1
Sample Student Report *(continued)*

Costs

This proposal will require one working day for a security worker to perform the proper analysis. It would take about half a day for the security worker to install the trivial alarm system recommended by Schechter. The analysis and implementation will begin as soon as you give the go-ahead. More information about the cost is in Table 1.

Table 1 Tasks and Estimated Costs

Task	Estimated Cost
Gathering data	Approximately 4 ($160) working hours
Performing analysis	Approximately 2 ($80) working hours
Evaluating the findings	Approximately 2 ($80) working hours
Implementing a trivial alarm system	Approximately 4 ($160) working hours and $5,000 for computer security enterprise alarm system
Implementing security solutions to the liking of ABC Corp.	Variable

From Table 1, we can see that it would cost approximately $320 for ABC to perform the RROI analysis. Because of the large potential cost savings and optimization that this method offers, I strongly recommend that ABC follow this path. After the analysis has been performed, ABC could opt for a computer enterprise alarm system or other security solutions according to the strategy recommended by the methodology. ABC Corp. will be able to calculate the exact amount of money it has saved by calculating its new observed damages and bypass rates.

Conclusion

I am enthusiastic about establishing the much-needed computer security methodology discussed here. After an initial low analysis cost, ABC Corp. will be able to optimize its spending for security and gain an invaluable return on its money. This proposal can take place as soon as you are ready. Please contact me for further discussion when you are ready.

Figure 4.1
Sample Student Report *(continued)*

References

Arora, A., Hall, D., Ariel Pinto, C., Ramsey, D., & Telang, R. (November–December 2004). Measuring the risk-based value of IT security solutions. Retrieved January 23, 2007, from *IEEE Computer Society Digital Library*, *http://csdl.computer.org/dl/mags/it/2004/06/f6035.pdf.*

CERT. (January 16, 2007). CERT/CC statistics 1988–2006. Retrieved January 29, 2007, from CERT Web site: *http://www.cert.org/stats/cert_stats.html.*

Schechter, S. (January/February 2005). Toward econometric models of the security risk from remote attacks. Retrieved January 23, 2007, from *IEEE Computer Society Digital Library*, *http://csdl.computer.org/dl/mags/sp/2005/01/j1040.pdf.*

Strassmann, P. (November 2006). 3 Metrics to gauge security spending. *Baseline*, 66, 32. Retrieved January 23, 2007, from *EBSCOhost Database*, *http://find.galegroup.com/ips/infomark.do ???&contentSet=IAC-Documents&type=retrieve&tabID=T003&prodId=IPS&docId=A105884245&source=gale&userGroupName=iulib_iupui&version=1.0.*

Vijayan, J. (July 21, 2003). IT security short on funding: Building a case will require alternative to usual ROI techniques. (Ernst and Young International survey). *Computerworld*, 37, 1(2). Retrieved January 23, 2007, from *Computer Database* via Thomson Gale, *http://find.galegroup.com/ips/infomark.do?&content Set=IAC-Documents&type=retrieve&tabID=T003&prodId=IPS&docId=A105884245&source=gale&userGroupName=iulib_iupui&version=1.0.*

Appendix A: Total Vulnerabilities Reported

Figure 4.1 Sample Student Report *(continued)*

Table 2 Year and Total Vulnerabilities Reported

Year	Vulnerabilities
1995	171
1996	345
1997	311
1998	262
1999	417
2000	1090
2001	2437
2002	4129
2003	3784
2004	3780
2005	5990
2006	8064

(Source: CERT, 2007, p.1)

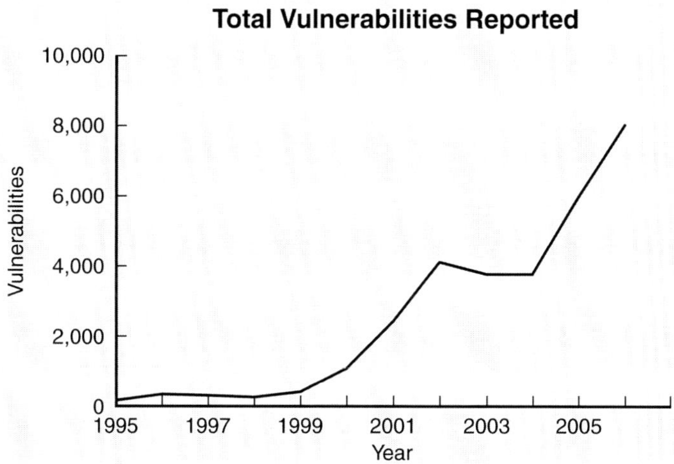

Figure 4
Total Vulnerabilities Reported
(Source: CERT, 2007, p. 1)

Figure 4.1
Sample Student Report *(continued)*

Appendix B: Total Intrusions Reported

Table 3 Year and Total Incidents Reported

Year	Incidents
1988	6
1989	132
1990	252
1991	406
1992	773
1993	1334
1994	2340
1995	2412
1996	2573
1997	2134
1998	3734
1999	9859
2000	21756
2001	52658
2002	82094
2003	137529

(Source: CERT, 2007, p.1)

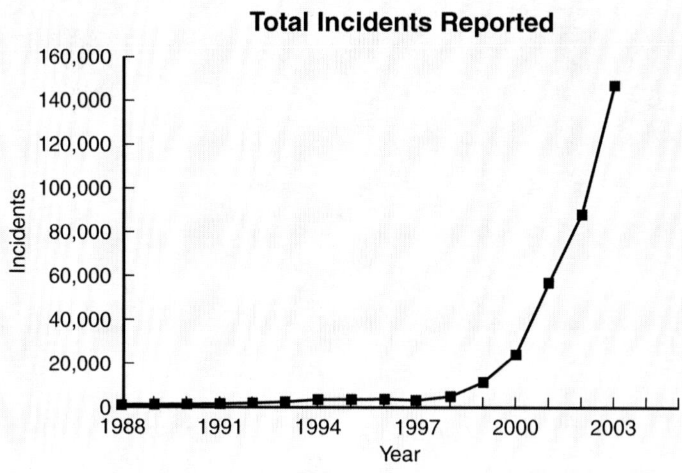

Figure 5

Total Incidents Reported
(Source: CERT, 2007, p. 1)

Figure 4.2 Sample Student Report—with annotation Comparison (with Recommendation)

Bligh Consulting

To: William Bligh, Owner
From: Kenneth Jenkins
Date: October 16, 2006
Re: File Storage Alternative

Comment [cnc1]: First major heading (signals Beginning.)

Introductory Summary

Comment [cnc2]: Orienting sentence (reminds audience of previous interaction on this subject.)

As requested, I have researched three options to replace your current usage of 1.44 MB floppy disks for your personal file storage. The three options you expressed interest in were evaluated: USB flash drives, CD-RWs, and zip disks. Research was done by going online to review technical data and pricing. After talking with you about your usage habits and storage needs, I have based my recommendation on the following criteria:

- Storage Capacity
- Portability (physical size)
- Cost
- Durability
- Practicality (ease of use)

Comment [cnc3]: Guiding verb: recommend.

Based on the criteria listed, a USB flash drive was found to be the best option for your data storage needs. I therefore recommend using a USB flash drive for your data storage needs because they are the most practical, portable, and durable of the three options.

Comment [cnc4]: Second major heading (signals Middle.)

Discussion

Since purchasing a new laptop that did not facilitate the use of 1.44 MB floppy disks, you are in need of a data storage alternative. I was charged to research three options that you were the most interested in. Per your request, I have gone online to research the three options based on the aforementioned criteria.

The three options under consideration were: USB flash drives, CD-RWs, and zip disks. Each option has its own "pros" and "cons." The three (3) options were evaluated on the following five (5) key factors: storage capacity, portability, cost, durability, and practicality.

Comment [cnc5]: First subheading (signals first supporting point.)

Storage Capacity

Based on your needs, storage capacity was a concern. The use of multiple 1.44 MB floppies was becoming cumbersome as your personal files became more numerous and increased in size. All three of the options being considered offer considerably more storage capacity than a 1.44 MB floppy. USB flash drives offer a wide array of different capacities ranging from 512 MB to 4 Gigabytes; zip disk capacities come in three (3) sizes: 100 MB, 250 MB, and 750 MB. The storage capacity for CD-RWs is 700 MB.

Figure 4.2
Sample Student Report—wtih annotation Comparison (with Recommendation) *(continued)*

Comment [cnc6]: Second subheading (signals second supporting point.)

Portability

Since you are away from the office on numerous occasions, the new storage method needs to be very portable. During my research, I found the USB flash drive to be the most portable due to its small size. CD-RWs and zip disks are portable as well, but are a little bit larger than a USB flash drive and require protective cases when not in use. Zip disks will also require an external zip drive since your laptop does not have a zip drive, thus making them less portable.

Comment [cnc7]: Third subheading (Signals third supporting point.)

Cost

Being the owner of a small business, one has to count every penny. Cost is a major consideration in the choice of the ideal candidate. I found CD-RWs to be the cheapest of the three options. The most expensive option would be the zip disk. Even though the USB flash drive has the higher initial cost, the zip disk would require you to purchase an external 100 MB, 250 MB, or 750 MB zip drive, which would add an additional $100 to $200 to the overall cost.

Comment [cnc8]: Fourth subheading (signals fourth supporting point.)

Durability

Since you sometimes find yourself working out in the field and are always on the go, another consideration is the durability of the ideal candidate. Imagine you are with a client and you cannot access your data because the CD-RW is scratched. The ideal candidate must be very durable to avoid such incidents.

I found the USB flash drive to be the most durable. For one, it is not a mechanical device. Mechanical devices are prone to eventually wearing down. Second, you can save your files to the USB flash drive as many times as you need to. Third, the USB flash drive is contained within its own protective covering. Fourth, the average lifespan of a USB flash drive is approximately ten (10) years.

CD-RWs, on the other hand, are not as durable. CD-RWs can be written to approximately 1,000 times, can be scratched easily, and need to be placed in a protective case when not in use. The expected lifespan for CD-RWs is approximately two (2) to five (5) years. Because of these factors, I would strongly discourage the use of CD-RWs for storing your files if you plan to use them outside of the office environment.

Zip disks are more durable than CD-RWs, but not as much as USB flash drive. Zip disks have a protective outer shell, but it is recommended to place them in a protective case when not in use. They are prone to being crushed or bent. Zip disks and zip drives are also prone to mechanical failures. The expected lifespan for zip disks ranges from two (2) to ten (10) years.

Comment [cnc9]: Fifth subheading (signals fifth supporting point.)

Practicality

During our meeting, you expressed an interest in something that could be accessed quickly and easily while you were on the move.

I found the USB flash drive to be the ideal candidate because there is no additional software, equipment, or computer "know how" to operate. All you need to do is simply plug the USB flash drive into an available USB port on your laptop or desktop and you will have access to your files.

CD—RWs require additional software and computer "know how" to save files to the disk. Zip disks require an external zip drive that must be carried around and attached to your laptop through a USB or port.

Conclusion

While all three options are viable, USB flash drives offer the best alternative to your current use of 1.44 MB floppy disks for your storage needs. USB flash drives top the list of criteria except for cost. I recommend purchasing a USB flash drive because they are the most portable, dependable, and easy to use of the three (3) options. Zip disks require additional equipment that both your laptop and desktop computers don't possess. As far as CD-RWs are concerned, durability is the major disadvantage for this storage medium. They are easily scratched and their lifespan is only two (2) to five (5) years.

If you need further assistance with your decision making on which storage device to use, I will be more than happy to meet with you. If you have any further questions, please contact me.

Once you have made your decision, allow me to help with the purchasing and orientating you on how to use the option that you have chosen.

Figure 4.2 Sample Student Report—with annotation Comparison (with Recommendation) *(continued)*

Comment [cnc10]: Third main heading (signals End.)

Comment [cnc11]: Natural conclusion drawn from previous information.

Comment [cnc12]: Reiteration of recommendation. Third main heading (signals End.)

Comment [cnc13]: Ends with offer to do more, expression of good will.

Chapter Summary

Using the suggested techniques and strategies outlined in this chapter will help you gain control of both the process and the product of workplace informal reports. Looking at the report from a logical perspective, you simply need to think of the requirements of the beginning, middle, and end. All have their own purposes and structures. Good preparation will generate the needed details and aid the design. The strategic and careful use of headings, subheadings, bulleted or numbered lists, and white space will also help create a clear and accessible document.

End of Chapter Activities

ACTIVITY 1 Headings Practice

Bring a workplace document to class that doesn't include sensitive information. If you are not currently working or can't bring a document from work, find a document from another source. (Delete names to protect the innocent, if appropriate.) Form groups of three to five and analyze the documents each classmate brought. Are headings used? If so, are they effective? How would you revise to make the headings more effective in each of the documents? Be prepared to report your results to the class.

ACTIVITY 2 Executive Summary Practice

Form groups of three to five people. Read the following quoted piece and write an Executive Summary of the information.

Oral History Project Of the Senate Historical Office

About the Project

Since 1976 the Senate Historical Office has interviewed Senate officers, parliamentarians, clerks, police officers, chiefs of staff, reporters, photographers, Senate pages, and senators. These interviews cover the breadth of the 20th century and now the 21st century, and include a diverse group of personalities who witnessed events first-hand. Darrell St. Claire, Assistant Secretary of the Senate, offered reminiscences of senators from Huey Long to Lyndon Johnson. Ruth Young Watt, Chief Clerk to the Subcommittee on Investigations under Joseph McCarthy and "Scoop" Jackson, candidly described fellow staffers Roy Cohn and Robert Kennedy, and reminisced about witnesses such as Howard Hughes and Jimmy Hoffa. Jesse Nichols, clerk and librarian for the Finance Committee from 1937 to 1971, was the first African-American hired on the Senate's clerical staff. He spoke of the long, slow transition from a segregated city to an integrated workplace.

Senate historians interview those individuals who can offer a unique perspective on Senate history but may otherwise be missed by biographers, historians, and other scholars. Steeped in the literature and folkways of the Senate, the Senate historians also have ready access to Senate records.

The Historical Office has also received strong institutional support from the Senate and its members. In 1984, for instance, the Appropriations Committee chairman, Senator Mark Hatfield, literally ordered the retiring Secretary of the Senate William Hildenbrand to give an interview. Hatfield humorously threatened to suspend Hildenbrand's salary "until that oral history is either arranged or has been completed." Hildenbrand complied with the request.

Many of the earliest interviewees were men and women who began their Senate career when the Legislative Reorganization Act of 1946 mandated a larger nonpartisan and professional staff. They described the evolution of the Senate from a small community or "inner club," often closed and mysterious, to a complex, modern institution increasingly conducting its business in full public view.

The interviews follow a biographical format, beginning with the subject's family background, education, and pre-Senate work experience. They explore how Senate service began, and then trace those careers from senator to senator, office to office, and committee to committee, frequently making comparisons among these many settings and personalities. Senate historians interviewed the first three chiefs of staff for the Senate Foreign Relations Committee, covering the period of 1945 to 1977. The first, Francis Wilcox, remembered chairman Arthur Vandenberg stopping by his desk, cigar in hand, and engaging him in extensive discussions of foreign policy. Wilcox's successors, Carl Marcy and Pat Holt, carried the behind-the-scenes action to ninety-one-year-old Theodore Francis Green, maneuvered out of his chairmanship by majority leader Lyndon Johnson, and then to Green's successor, J. William Fulbright, who used his position as chairman to battle President Lyndon Johnson over the Vietnam War. Reporter of Debates Francis Attig observed the change in Senate oratory from the 1950s to the 1970s. Others commented on the influence of social and cultural changes on the work environment.

Transcripts of the interviews are deposited in the Library of Congress, the National Archives, and the appropriate presidential libraries and senatorial manuscript collections. Some interviews are now available on the U.S. Senate Web site.

Source:

http://www.senate.gov/pagelayout/history/g_three_sections_with_teasers/oralhistory.htm

CHAPTER 5

Types of Informal Reports

Overview

This chapter discusses the various kinds of management reports you are likely to encounter in the workplace. As you read through it, think about similar writing challenges you have already encountered.

The chapter concentrates on three major types of workplace reports: the comparison report, the problem analysis/solution report, and the recommendation report. The order in which we discuss them reflects to some extent their increasing complexity. The comparison report relies on the fairly simple concept of comparing items, of discerning their similarities and differences. The problem analysis report challenges you to break a problem apart into its contributing factors and then to evaluate those factors. The act of analysis is a crucial one in critical thinking. Finally, the recommendation report asks you to apply what you have learned in the previous reports to create a persuasive recommendation based on your interpretation of the data.

Keep in mind that although we present these report types separately and discretely, in reality, they share a lot of common characteristics. Similar to learning scales and chords before you play a musical piece, the early reports build your skills for the later ones. You will find that ultimately you will often combine several approaches into one activity. For example, you may analyze a situation in order to arrive at the criteria for comparison, which will lead to your recommendation. The intention of this chapter, then, is to provide the overall concepts that will serve as a foundation for your growing communication skills.

Comparison Report

Comparison is such an integral part of our everyday existence that we rarely think about how important the simple act of comparison is. Whenever you decide between two or more choices, you're using comparison. In the workplace, you will be asked at some point to judge two or more products, processes, activities, ideas, concepts. When you do, you'll want to have a systematic way of deciding between the options.

Purpose =
To inform and to persuade

Using comparison as a way of sorting choices is a common cognitive exercise. When you go to a restaurant, for example, you look at the menu in order to make your selection. Depending on your circumstances, you order a big burger with fries, a cup of soup and a salad, a gyro sandwich, or vegetarian chili. What mental process, however unconscious, aids you in making your choice?

Well, for one, maybe you have only $5.00 in your pocket, and you're not using credit cards these days. Second, you skipped breakfast and are very hungry, and you know you won't have dinner until late this evening. Although you've tried to decrease the amount of red meat you have been consuming, you haven't given up meat entirely, so it is still a possibility. All these factors contribute to your decision.

These factors are also called *criteria:* standards on which to base your decision. In the previous example, you could identify the criteria as (1) available funds, (2) your hunger factor, and (3) meatless content. Once identified, the criteria can be judged either *essential* or *desirable*. *Essential* criteria are those that are absolutely necessary, "deal breakers," in other words. *Desirable* criteria are certainly preferred, but negotiable. Then you can chart your decision. See Table 5.1.

Table 5.1 Charting Your Decision

Criteria	Burger & Fries	Soup & Salad	Gyro	Vegetarian Chili
Under $5 (essential)	$6.25 (No)	$3.89 (Yes)	$4.50 (Yes)	$4.25 (Yes)
Filling (essential)	Yes	No	Yes	Yes
Meatless (desirable)	No	Yes	No	Yes

Looking at the possibilities, you immediately find a solution. On the basis of how your choices match up against the criteria, the obvious choice is the vegetarian chili. Turns out, however, that a gyro simply sounds better today, and so you decide on it. The essential criterion of under $5 did predetermine that the burger and fries were out. The desire to be full eliminated the soup and salad. But the meatless criterion was one you decided to forgo.

Okay, so you don't take out a sheet of paper to make a chart. But your thinking process is very similar to what is outlined here.

Criterion = Singular form of the word

Criteria = Plural form of the word

Making business choices often works the same way. For example, you may work for a company that buys maintenance services for technical equipment. The time has come either to renew the contract with the current provider or to choose another one. How do you go about that?

The following outlines a process you can use to make an informed decision.

GATHER DATA

Do your research. Talk to your supervisor to find out what her needs are, talk to the finance department, inventory your own experience and knowledge. Look at your company's needs carefully before deciding on that final list of criteria.

IDENTIFY AND DEFINE CRITERIA

Now, identify the criteria based on your research. Note that these criteria are not universal. For example, a hospital choosing a vendor may require 24-hour/7-day week availability. However, a small business office may require only standard business hours: Monday through Friday, 8 to 5.

When you choose the criteria on which to base your comparison, make sure that you not only *name* the criteria, but that you also *define* them. Some criteria may seem to be workable but are, in fact, simply not. For example, one common criterion is, logically enough, *price* or *lowest price;* but what exactly do those words mean? If nothing else, generally you have a maximum amount you want to spend. So even if you are assuming that the lowest price is the most desirable price, the lowest price cannot exceed your maximum price. Thus, if you want to buy a used hybrid car, and you have only $12,000 to spend, and the cheapest of your choices—the '09 Prius—costs $15,600, you cannot buy the '09 Prius. So the *cheapest price* is not the determiner; the *cheapest price under $12,000* is the criterion.

The same comments apply to abstractions like quality, taste, or attractiveness. Keep in mind, aesthetic judgments are subjective, and different people have different preferences and tastes. Unless you can define in the context of the situation what you mean by *quality*, it has virtually no meaning. For example, if you are using *quality* as a criterion for selecting service providers, *quality* may be defined in terms of guaranteed response times and technician certifications; in terms of selecting bed sheets, *quality* may be defined in terms of fabric content and thread count. In any case, the comparison will not work until the criteria are named and defined.

Remember: When it comes to criteria, one size does not fit all.

RESEARCH OPTIONS

Once you have determined and defined the criteria, based (again!) on the needs of your particular situation, you are ready to research the options. You may be asking, how do I choose what options to compare? Even this decision is based on a set of criteria—right? Did your boss ask you to compare X, Y, and Z? Do the options fall into the right price category? Do the options best fit the criteria you've chosen? Do the options compare logically on some other level? For whatever reason—hopefully a logical and ethical one—you have chosen to compare options X, Y, and Z. Now it is time to research these options.

Remembering that you must decide on one of the options or that the information you gather must help your reader decide on one of the options should be motivation enough to persuade you to do a thorough job of gathering relevant data about the options. When you have gathered your data, you should decipher systematically what you have found. That is where a Comparison Log comes in.

COMPARISON LOG

A Comparison Log helps you organize the data you have gathered and ultimately helps you decide on which option best meets your criteria. The log gives you an effective visual that displays the data so you can compare and contrast how options are similar and different, as well as how they compare to the criteria you have set. Although it does not draw conclusions for you, it allows you to see the data in aggregate format, thus simplifying your task of analyzing the data and drawing conclusions. The log also makes it much easier for you or your reader to recommend one of the options. Make a comparison log and chart the data you have gathered. See Figure 5.1 for a student sample comparison log and Figures 5.2 and 5.3 for examples of effective uses of such a log in the workplace.

Figure 5.1 Sample Student Comparison Log (Eden Dorsett, student)

Comparison of Compact Utility Tractors

IDEAL (criteria)	REAL (evidence)			JUDGMENT (difference)		
	A	B	C	A	B	C
Priced Under $30,000 (Cheaper Is Better)	Yes $26,500	Yes $24,639	Yes $21,992	+	+	+
Option for Locking Rear Differential	No	Yes	Yes	–	+	+
Option for Automatic 4-Wheel Drive Engagement	No	No	Yes	–	–	+
Sloped Front Hood for Good Visibility	No	Yes	Yes	–	+	+
3-Pt Hitch Lift Capacity at 24" behind Lift Point	2,756 lbs	2,500 lbs	2,330 lbs	+	–	–
Option for SST[a] or HST[b]	Yes	Yes	Yes	+	+	+
Option for Mid-Mount PTO[c]	Yes	Yes	Yes	+	+	+
Standard 2-Yr Warranty	Yes	Yes	Yes	+	+	+
Total Number of Positive Marks				5	6	7

A = Kubota (L4240); B = John Deere (4120); C = New Holland (Boomer 3045)
[a]SST—Shuttle Shift Transmission
[b]HST—Hydrostatic Transmission
[c]PTO—Power Take Off

Typemock Products Comparison Matrix

	Ultimate	Isolator 2010	Isolator 2010 for Web Developers ASP.net	Isolator 2010 for SharePoint Developers
Powerful isolation engine	✔	✔	✔	✔
Intelligent test completion	✔	✔	✔	✔
Full documentation	✔	✔	✔	✔
Discoverable API	✔	✔	✔	✔
Change resistant tests	✔	✔	✔	✔
Recursive fakes	✔	✔	✔	✔
Dedicated support stuff	✔	✔	✔	✔
Community support in forums	✔	✔	✔	✔
Write SharePoint unit test	✔	✔	✔	✔
Write ASP.Net unit test	✔		✔	
Write ASP.Net MVC	✔	✔	✔	
Typemock Tracker Server	✔			
TestDriven.Net	✔			
Save	$350 per developer			
Price	$1,199	$799	$899	$249

Figure 5.2
Sample Professional Comparison Log

Source: http://site.typemock.com/comparison-matrix/

Key
↑ Fulfills Requirement
↓ Does Not Fulfill Requirement
↔ Might Fulfill Requirement

	OPTION A	OPTION B	OPTION C	OPTION D	OPTION E
	Make No Improvements and Eventually Discontinue Operation	Renovate Existing Facilities	Replace Facilities on Existing Site	Replace Facilities on Identified Site on IUPUI Campus	Replace Facilities on Unidentified Off-campus Location
Provides Value for the Community	↓	↓	↔	↑	↓
QUALITY OF CARE AND SAFETY					
Meets Current and Forecasted Needs of Indigent and Underserved	↓	↓	↔	↑	↓
Meets Hospital Standards for Room Size	↓	↓	↑	↑	↑
Meets Hospital Standards for Privacy	↓	↓	↑	↑	↑
Meets Hospital Standards for Access	↓	↓	↔	↑	↔
Meets New Hospital Current Standards for Air Quality	↓	↓	↑	↑	↑
Facilities for Best-Practice Standards for Quality Health Care Delivery	↓	↓	↑	↑	↑
Clinical Operation Efficiency	↓	↓	↔	↑	↓
Disruption to Operations During Implementation	↓	↓	↓	↑	↑
Patient Rooms ADA Compliant/Bariatric Inclusive	↓	↓	↑	↑	↑
Promotes Recruitment and Retention of Physicians	↓	↓	↔	↑	↓
Promotes Recruitment and Retention of Staff	↓	↓	↔	↑	↓
Supports Level 1 Trauma Center	↓	↓	↑	↑	↓
Supports Community's Disaster Planning and Response	↓	↓	↔	↑	↔
Supports Burn Center	↓	↔	↑	↑	↑
Support Senior Care	↓	↓	↔	↑	↓
Supports Women's Health	↓	↓	↔	↑	↓
Supports Specialty Clinic Services	↓	↓	↔	↑	↓
Supports HIV Care and Treatment	↓	↓	↓	↑	↓
FISCAL EFFECTIVENESS					
Construction Cost	↑	↓	↓	↑	↓
Facility Maintenance Costs	↓	↓	↓	↑	↑

Figure 5.3
Sample Professional Comparison Log

Operations Costs	↓	↓	↑	↑	↑
Financial Responsibility and Sustainability	↓	↓	↓	↑	↓
Local Economic Stimulus	↓	↑	↑	↑	↑
Disruption to Operations During Implementation	↓	↓	↓	↑	↑
Ability to Satisfy Programmatic Growth	↓	↓	↔	↑	↓
OPERATIONAL EFFECTIVENESS					
Efficient Use of Space	↓	↓	↔	↑	↑
Reduction in Energy Use	↓	↓	↑	↑	↑
Time to Implementation	↑	↓	↔	↑	↓
Abundant and Convenient Parking	↓	↓	↔	↔	↔
Clinical Operation Efficiency	↓	↓	↔	↑	↓
Supports Increase in Productivity	↓	↓	↔	↑	↓
Non-Clinical Operation Efficiency	↓	↓	↔	↑	↑
Access to Public Transit System	↑	↑	↑	↑	↔
Supports Link to IU School of Medicine	↓	↑	↑	↑	↓
Supports Link to Clarian	↓	↑	↑	↑	↔
Supports Link to RL Roudebush VA Medical Center	↓	↑	↑	↑	↓
Disruption to Operations During Implementation	↓	↓	↓	↑	↑
Facility Promotes Recruitment and Retention of Physicians	↓	↓	↔	↑	↓
Facility Promotes Recruitment and Retention of Staff	↓	↓	↔	↑	↔
Patient Transport	↓	↓	↔	↑	↑
Supply Transport	↓	↓	↔	↑	↑
Ability to Satisfy Programmatic Growth	↓	↓	↔	↑	↓
PATIENT ACCESS AND WAYFINDING					
Meets Hospital Standards for Access	↓	↓	↑	↑	↔
Clear and Friendly Wayfinding	↓	↑	↑	↑	↑
Pedestrian Circulation	↓	↓	↑	↑	↑
Abundant and Convenient Parking	↓	↓	↔	↔	↔

Figure 5.3
Sample Professional Comparison Log
(*Continued*)

Access to Public Transit System	↑	↑	↑	↑	↔
Traffic Circulation	↓	↓	↓	↓	↔
Disruption to Operations During Implementation	↓	↓	↓	↑	↑
Separation of Public, Service, Staff and Ambulance Traffic	↓	↓	↔	↔	↔
TEACHING, RESEARCH AND INNOVATION					
Supports Link to IU School of Medicine	↓	↓	↑	↑	↓
Promotes Education	↓	↓	↑	↔	↓
Promotes Recruitment and Retention of Physicians	↓	↓	↔	↑	↓
Promotes Recruitment and Retention of Staff	↓	↓	↔	↑	↓
Promotes Research	↓	↓	↑	↔	↓
SITE PLANNING					
Sustains Planned Growth and Pathways	↓	↓	↓	↑	↔
Retail Potential	↓	↓	↓	↔	↔
Access to Public Transit System	↑	↑	↑	↑	↔
Pedestrian Circulation	↓	↓	↓	↑	↔
Engages and Supports Neighbors	↓	↓	↓	↑	↑
Abundant and Convenient Parking	↓	↓	↔	↔	↔
Separation of Public, Service, Staff and Ambulance Traffic	↓	↓	↓	↔	↔
Proximity To Green Space	↓	↓	↔	↑	↔
Proximity To Indiana Greenways	↑	↑	↔	↑	↔
Zones of Growth	↓	↓	↓	↑	↔
TIME TO COMPLETION					
Overall Time to Completion	↓	↓	↓	↑	↓

Figure 5.3
Sample Professional Comparison Log
(*Continued*)

STEPS OF A COMPARISON REPORT

1. Identify the need. (Sometimes you will initiate the process; sometimes others will initiate it for you.)
2. Gather data from decision makers to determine criteria, when possible.
3. Add own criteria to the list, if appropriate.
4. Define the criteria as specifically as possible.
5. Narrow the list to the most important criteria to the decision makers.
6. Classify those criteria as desirable or essential.
7. Gather information to make your comparison.
   a. Choose potential candidates that would fit your criteria.
   b. Chart your results in a comparison log.
8. Evaluate your candidates based on the information gleaned from step 7.
9. Organize those factors in an appropriate order. You have two common choices:
   a. **By criterion:** Take your most important criterion; then report how each candidate measures against it. Then take the next criterion, and do the same. Consistently follow this pattern until all of the criteria have been discussed. Make sure you discuss the candidates in the same order each time. Organizing by criterion is usually preferable over organizing by candidate.
   b. **By candidate:** Discuss the first candidate, criterion by criterion. Then discuss the second candidate against all the criteria in the same order as you discussed the first candidate. Consistently follow this pattern until all of the candidates have been discussed.
10. Create your report around those details. You may choose to present your findings and conclusions so that your reader can make a decision, or you can choose to make a recommendation to your reader based on what you have found. That decision rests on your relationship with your reader and on the expectations she/he has for the report.

Problem Analysis Report

Purpose =
To inform

When you are faced with a problem that you want to fix, it is often too easy to jump to a solution. Sometimes, however, this solution works for a symptom, but not for the actual problem. You want to fix the problem, but you need to know exactly what the problem is. This is the *problem* part.

The *analysis* part is figuring out possible contributing factors to the problem. If you can discern the factors, you have a better chance of correcting them. Instead of trying to solve the entire problem globally, you divide it into units and look at the units one by one.

For example, let's say that you are an office manager for a small firm in charge of ordering supplies, paying vendors, and generally keeping the office's day-to-day operations running smoothly. You work within a monthly budget. For the past quarter, you have noticed expenses going up, and you are increasingly challenged to stay within the budget. Furthermore, in an effort to stay fiscally sound, you are not ordering as many supplies, and in a couple of instances, you have even run out of crucial materials.

So the overall problem is to keep the office running efficiently within the budget. But before you jump to a quick solution—"I need a bigger budget"—you want to investigate what has changed and why. How do you go about that? We recommend the following process.

IDENTIFY POTENTIAL CONTRIBUTING FACTORS

The first challenge in the process of analyzing a problem is to identify potential contributing factors.

In this example, you might begin by itemizing the budget that you oversee. Where does the money go? You might list office supplies such as paper, copier charges, pens, pencils, and paper clips. You look at utility charges: heat, electricity, water, telephone. You identify travel expenses such as airline tickets, hotels, food, and mileage. You include lease payments and insurance costs. In other words, you divide the overall concept of "expenses" into its component parts. So, *expenses* could be a contributing factor. (Other contributing factors might be that staff is being wasteful or that additional staff have been hired.)

Like the criteria you establish to conduct your comparison, the factors you choose to do your problem analysis must also be as quantifiable and as definable as possible. For instance, if you are positing that *morale* is a factor in decreased employee productivity, you need to think about how morale is defined and measured. Is poor morale reflected in an increase in absenteeism, comments received during exit interviews, or results revealed in an employee survey? If *poor communication* emerges as a potential factor in the problem you are analyzing, again, how do you define that abstraction? By closely defining the measurable characteristics of the factors, you can later determine if the solutions you implement have worked. The important concept here is to dig into the factors you are exploring, not just to stay on the surface of them.

MEASURE THEIR EFFECTS

The next challenge is to measure the effects of the contributing factors. Let's go back to the earlier example of the office manager budget problem. Taking the expense categories you have identified, you study the related charges that have occurred over time. Charting out the expenses may provide you with the information you need to confirm your suspicions. You eliminate and quantify. For example, on the one hand you find that the actual cost of office supplies has not changed significantly. On the other hand, you may identify that travel expenses have taken a serious turn upwards: Airline expenditures have risen significantly since last year.

You can then take the analysis to the next step: Is the increase in airline expenditures due to the cost of tickets, or are more employees flying these days?

Now your reader is in a better position to determine a solution to the problem. By breaking the problem into its potential parts, gathering data, measuring the effects, and presenting your findings in an organized report, you are enabling your reader to be able to make a reasoned and informed judgment. Presenting this intelligent analysis to the decision maker, you make it much more likely for him or her to fix what needs to be fixed, rather than to fix what is not broken.

STEPS OF A PROBLEM ANALYSIS REPORT

1. Identify the problem. (Sometimes you will initiate the process; sometimes others will initiate it for you.)
2. List what you see as the reasons (contributing factors) for the problem.
3. Gather information to test your reasons.
   a. Collect quantitative data that test your reasons.
   b. Conduct a survey.
   c. Interview people.
   d. Do secondary research on the topic.
4. Based on the information gleaned from step 3, reevaluate your analysis and pinpoint what you see as the relevant contributing factors.
5. Organize those factors in an appropriate order based on your research.
6. Create your report around those details.
7. The reader will make a decision or pass the problem analysis to someone who will.

Problem/Solution Report

Purpose =
To inform and to persuade

Problem/solution follows the same principles as problem analysis, in that its purpose is to get to the causes of the problem. The difference comes in follow through. Unlike a problem analysis report, which leaves the solution up to the reader, a problem/solution report suggests a solution based on evidence.

The same principles just discussed on pp. 86–87 define the initial stages of the problem/solution process. Your initial steps are to identify the potential factors and measure their effects. The next step is to propose a solution.

IDENTIFY POTENTIAL CONTRIBUTING FACTORS

The first challenge in the process is to identify potential contributing factors. See Page 86.

MEASURE THEIR EFFECTS

The second challenge is to measure their effects. See Page 87.

PROPOSE A SOLUTION

The third challenge is to propose changes that will lead to a solution to the problem.

In the example just discussed, you may suggest that each department scrutinize its travel expenses to make sure that they are really necessary. By reducing those expenses by 10 percent, you may not need to increase the monthly budget at all. To make your case, you need to present your analysis and data as thoroughly and as convincingly as you can. You carefully explain your analysis, review the research you conducted to gather your data, and detail the thought processes that went into your conclusions. You want your reader not only to understand your solution, but also to agree with you about its merits.

Thus, the purpose of a problem/solution report differs from that of a problem analysis in that you are attempting to persuade, not merely to inform. Although both approaches use the same strategy to extract information, they differ in how they handle that information.

STEPS OF A PROBLEM/SOLUTION REPORT

1. Identify the problem. (Sometimes you will initiate the process; sometimes others will initiate it for you.)
2. List what you see as the reasons (contributing factors) for the problem.
3. Gather information to test your reasons:
   a. Collect quantitative data that test your reasons.
   b. Conduct a survey.
   c. Interview people.
   d. Do secondary research on the topic.
4. Based on the information gleaned from step 3, reevaluate your analysis and pinpoint what you see as relevant contributing factors.
5. Organize those factors in an appropriate order, e.g., most important to least important.
6. Determine what could be done to correct the situation.
7. Build your case around those conclusions, explaining your reasoning and backing up your conclusions with the evidence you have collected.
8. Propose a solution.
9. Write your report around your conclusions.

Recommendation Report

Purpose =
To inform and to persuade

We are all confronted with making many choices every day, from what we are going to wear to work to how many hours to donate to the calculus assignment to what movie to see or song to download. Sometimes just making the decision seems itself overwhelming, and we ask someone we know to *recommend* a solution. We want that person to gather the data and draw the conclusions and suggest the choice for us.

That process is exactly what drives the recommendation report. Your manager may have a lot of different projects and responsibilities, and he is not able to do all the research on every single thing himself, so he asks you to do it. In the scenario we previously discussed under comparison, for example, he may be tasked with choosing the new maintenance service provider. Instead of doing the research himself, he asks you to do the investigating, and then asks you to suggest what provider is best for your company.

The process of doing the research is very similar to the process used in either comparison or problem analysis (or a hybrid of both, as most real world situations are). The major distinction in the recommendation report is that its purpose is to persuade, to convince your reader that you have gathered good information, have done a thorough analysis of it, and are recommending the best solution. In other words, you have interpreted the data and are offering an informed recommendation. So the focus here is on presenting data that support your recommendation.

The situation just described happens when your manager assigns the project to you, in other words, has solicited your input. In some cases, you may yourself want to instigate a change. Perhaps you see a procedure or a policy in your workplace that you think needs to be changed. Perhaps you have identified a new product that your office should have, and you want to persuade your manager to purchase it. The following process will help you prepare a sound recommendation report.

GATHER DATA

To go about creating your report, you start with doing your research. You must collect your facts and figures before you can start interpreting them! Let's return, for example, to the scenario where you had the job of investigating service providers. You do your homework: Determine what your company's specific needs are; articulate the criteria you are going to use to analyze the data you collect; and ensure you understand the budgetary constraints. (You can gather this information in a number of ways: talking to the people who use the services; interviewing the finance person in charge of the budget; calling providers; studying providers' Web sites.)

EVALUATE DATA

Still using the vendor scenario, once you have gathered your information and done your analysis, you then choose which vendor would best suit your company's needs by weighing all the options against the criteria you've selected.

Introduction—anticipate no resistance and recommend up front

Introduction—anticipate resistance and make case; recommend later

MAKE RECOMMENDATION

You recommend that company. Your recommendation has credibility because you have done your homework and because you have facts that support your recommendation.

Your report will begin with an introduction including a purpose statement: to recommend a particular course of action. This approach is especially common if your manager has requested the information and recommendation. You may, however, be anticipating some resistance from your manager if you initiated the investigation. In that case, you will want to establish your credibility and make your argument first, delaying your recommendation until you have made your case. In either case, you focus on the information that supports your recommendation.

To present your case with maximum effectiveness, you have some strategic decisions to make. The following subsections offer some tips for fashioning an effective persuasive report.

Order of Information

The order in which you present your information is a good place to start thinking about your strategy. If you start with your recommended choice first, pointing out its virtues and then supplying the information about possible alternatives afterwards, that simple order may underscore your option as (literally) number 1. Alternatively, you may choose to build to the strength of your choice by discussing the other alternatives first, leading to the superiority of your choice as the final idea, the last thought the reader encounters. Which choice is better? Predictably, the answer to that question is, "It depends"—on your audience, on the context, on your evaluation of a good strategy.

For example, with a "get to the point" reader, you may choose to go right for the most substantive point; with a more "step-by-step" reader, building from weaker to stronger points may well be the strategic route to take. If the situation you are addressing is itself complex, starting with the smaller pieces that lead to the larger may make the most sense, whereas with a less complicated problem, centering the discussion on the major issue may be logical. Thus, we are back to "It depends."

Word Choice and Sentence Structure

We have previously alluded to word choice and sentence structure as important components of good technical communication, and their importance becomes increasingly clear and pertinent in this discussion on persuasion.

Using overtly comparative language—as long as it is honest and ethical—is one strategy. Emphasizing that your choice has a superior warranty or more horsepower than the others or the best ranking of the three in *Consumers' Report* reinforces your evaluation and recommendation. Instead of letting the facts speak for themselves, you take an active role in pointing out the advantages of your choice.

As you present your findings, use evaluative words to guide your reader to your point of view. Point out how you perceive the data you have uncovered: Instead of offering the statistic "47% of respondents to the survey answered 'very satisfied,'" you comment, "a *convincing* 47% of respondents. . . ." You can use contrastive language to drive your point home as well: "Whereas a *modest* 18% of respondents answered 'never again,' an *impressive* 47% checked 'very satisfied.'" Notice you have not changed the facts, but rather have presented your interpretation along with them.

Likewise, using strategic subordination can deemphasize any superior qualities the other candidate may have. Think of this idea as the "Yes, but" approach. Acknowledge reservations while dismissing them. In the following sentence, you can see how to apply the "Yes, but" approach: "Although candidates B and C are admittedly less expensive than candidate A, candidate A's superior mileage statistics and extended warranty quickly cut into that difference." Another way of presenting this information is, "If we look only at the initial price, candidate A appears to be a problem, but a closer examination of mileage statistics and extended warranty makes that conclusion less persuasive." Anticipating the opposing point of view does not mean being defeated by it; on the contrary, it gives you the opportunity to dismiss it before it takes root.

How deeply you go into the other side depends on the context, but conventional wisdom advises to be careful of not covering the other side too robustly, for fear of giving too much of a platform to a different point of view. Notice that often the acknowledgment of the other side appears in a subordinate clause, which even grammatically represents the superiority of your point of view.

Table 5.2 offers some word choices you might consider to add emphasis. You may also consult Figures 5.6 and 5.10 for examples of how previous students have effectively used the techniques described in the chapter.

Table 5.2 Words to Add Emphasis

Adjectives	Adverbs	Transitions	"Yes, but . . ."
Good, better, best	Very, too	However	Although
Less, least	Decidedly	Consequently	While
Superior/inferior	Almost, approximately	Furthermore, also	In spite of
Dramatic, compelling	Clearly	In contrast to	Granted
Insufficient, delinquent	Increasingly/decreasingly	Likewise	Even though
Convincing/unconvincing	Similarly	Not only . . . but also	If
Strong/weak	Critically		Nonetheless
Important/unimportant	Predictably		

STEPS OF A RECOMMENDATION REPORT

1. Identify the need. (Sometimes you will initiate the process; sometimes others will initiate it for you.)
2. Determine what information you need to gather to research your alternatives.
3. Gather that information.
4. Evaluate the information.
5. Decide which option you will recommend.
6. Create your report around those details that support your recommendation.

Progress Report

Purpose =
To inform and to persuade

Many of the reports discussed so far focus on a kind of one-stop-shopping approach, in that they concentrate on creating a single report to resolve a single workplace issue. You may find, however, that you are assigned to ongoing projects that may span several weeks to complete and that your supervisor or manager, or even your client, will expect periodic reports that lay out the status of the project. Called progress reports, these documents play an important role in keeping projects organized and focused, ensuring the project contributors stay on point, and identifying early any potential problems or delays. In addition to the written document—or in place of it—you may be asked to orally report on the progress of the project. Many of the same principles apply.

As with any other report, your first step is to discern what your audience needs to know. What are his priorities? You divide your project into component parts so that you can track and report on them separately. You create a project schedule. You identify the key people for each of the functions. You may use a project management software tool to automate all these activities. The progress report is a narrative that quickly summarizes the following:

- What is to be done
- What is being done
- What has been done
- What remains to be done
- Who is doing what
- When tasks are due
- What monies have been spent
- What complications have arisen or may arise

Generally, progress reports are expected at regular intervals, and they are approached consistently each time. In other words, the information is presented in the same order and under consistent headings each time.

STEPS OF A PROGRESS REPORT

1. Identify the project.
2. List the major categories of the project that need to be tracked.
3. Identify the major contributors of the project.
4. Set the project schedule.
5. Organize those factors in an appropriate order.
6. Create your report around those details.

Lab Report

Purpose =
To inform and to persuade

Lab reports are common in engineering and scientific fields and may also be used in many other fields. Here we will discuss a general format for lab reports that you can adapt to various situations. Lab reports have two basic functions:

1. To report the results of an experiment or an activity you've performed
2. To report the significance of those results

When writing a lab report, you need to report more than data; you need to show that you understand the meaning of the data. Communicating precisely and clearly is essential.

Although your work or school environment will determine what specifics go in your lab report, here are typical parts: title page, abstract, introduction, materials, methods, results, discussion, conclusion, references, appendix.

TITLE PAGE

The title page should include (1) the title of the project, experiment, activity, (2) your name (and those of your lab partners, if a group lab report), and (3) the date. The title needs to be succinct and accurately name the content of the project, experiment, or activity.

ABSTRACT (OPTIONAL)

For longer lab reports, including an abstract is a good idea. The abstract allows the reader to decide if she or he wants to read the full lab report. Conciseness is key in an abstract; a typical length is approximately one hundred words. An abstract summarizes the purpose, the data presented, and the major conclusions.

INTRODUCTION (OBJECTIVE)

The introduction states the subject; establishes the purpose, objective, or hypothesis of the project, experiment, or activity; and establishes the context by giving pertinent background.

MATERIALS (PHYSICAL PART DESCRIPTION) AND METHODS OR PROCEDURES

This section describes any materials or equipment used in the project, experiment, or activity, as well as the methods or procedures used. You'll walk a fine line here between giving enough information so that the reader can understand the experiment or activity and overwhelming him with too much information. Drawings are typical in this section.

RESULTS (EXPECTED AND OBSERVED RESULTS)

This section summarizes the results of the project, experiment, or activity. It will probably cover both the expected and observed results. Data are shared in graphs, drawings, photographs, tables, and figures, for example.

DISCUSSION

Your interpretation of the results goes in the discussion section. What is the significance of your results from a practical viewpoint? What is the significance from the industry's viewpoint? How do your results relate to existing theory and knowledge? Here, you'll discuss the logic behind your acceptance or rejection of the hypothesis and any future research needed.

CONCLUSION (SUMMARY OF RESULTS)

Some lab reports don't include a conclusion. But if you include one, a conclusion section ties everything together and brings closure to your lab report.

REFERENCES

Since more than likely you've included supporting literature in your discussion section, you'll need to include a references section that lists all sources (articles, books, and individual interviews, for example) cited in the lab report. Sources should be alphabetized; the citation style should conform to the style required by your workplace or course instructor.

APPENDIX

For some lab reports, you'll want to include additional information that doesn't fit in the report but adds background to help the reader understand your project, experiment, or activity. These items can be placed in an appendix. Refer to the appendix in the body of the lab report.

Whatever format you use to write your lab report, using principles of good writing is a must: well organized units, clear and concise sentences, accurate and appropriate language, standard grammar and mechanics. Although, in general, we strongly advise using first person and the active voice, in lab reports, the third person and passive voice are conventions often used in the procedures and results sections.

STEPS OF A LAB REPORT

1. Identify the project, experiment, or activity.
2. Identify pertinent background information necessary for the reader to understand the project, experiment, or activity.
   a. Locate previous research.
   b. Identify any terms that need to be defined.
   c. Establish the formulas you are using.
3. State hypothesis of the project, experiment, or activity.
4. Determine materials or equipment to be used.
5. List procedures.
6. Describe the results (expected and observed) of the project, experiment, or activity.
7. Discuss the significance of the results.
8. Suggest future research ideas.
9. List references used in the report.
10. Decide what needs to be depicted visually.
11. Identify any materials that need to go in an appendix.
12. Create your report.

Sample Student Reports

As a final aid, we have included nine sample reports, all taken from our former students (see Figures 5.4 to 5.12). Read through them and think about how they reflect the advice in this chapter. In some cases, we have noted important pertinent details for you to consider.

Figure 5.4
Student Report, Comparison (with recommendation)

Date: October 9, 2006
To: John Smith, Supervisor of Research and Development
From: Jason Brugh, Supervisor of Engineering
Subject: Evaluation of Thermometers for Laboratory Use

Introductory Summary

As per your request, the Engineering Department has evaluated four different styles of laboratory thermometers that are to be used to monitor freezers and refrigerators that hold product samples. After evaluating several key factors, the Engineering Department has concluded that the digital thermometer is the best solution for your application. In this comparison report, we will give you the supporting details for our decision.

Discussion

In the evaluation to find the appropriate thermometer for your application, we researched the specifications to four different styles of thermometers: glass, infrared, bimetal, and digital. To compare each of the types, we used ideal criteria which included build characteristics, accuracy, expense, and life expectancy. The following is a summary of how each was researched and how the individual model compared to the ideal criteria.

Glass

While the glass mercury thermometer was an inexpensive choice and had a life expectancy of more than two years, the accuracy of the thermometer was not within the range of 0.1 degrees. Most importantly, glass and mercury were found to be unsafe for use in a laboratory setting and therefore did not meet the qualification for purchase. Because of the extreme temperatures found in the laboratory freezers and the constant use by the laboratory technicians, we feel the risk of mercury poisoning is too great should a glass thermometer be damaged during normal daily operation. Our data were provided by www.coleparmer.com, www.omega.com, and technical safety standards for mercury thermometer use. Sample pictures of glass mercury thermometers can be found in the Appendix of this report.

Infrared

The infrared thermometer was found to be safe to use in the laboratory, easy to use, rugged, accurate within 0.1 degrees, and expected to exceed the two-year life expectancy. However, the price of the unit made it not cost effective for us to purchase. Also, infrared thermometers rely on a surface for temperature measurement. The laboratory requirement was to measure air temperature in both freezers and refrigerators, making an infrared thermometer unusable for the application. We used www.omega.com/prodinfo/infraredthermometer.html and the "IR Principles" document provided by www.raytek.com. A sample picture of an infrared thermometer can be found in the Appendix.

Figure 5.4
Student Report, Comparison (with recommendation) *(continued)*

Bimetal

Bimetal thermometers use two different types of metal that create a voltage when they are heated or cooled. The bimetal thermometers were found to be the least likely candidate for use in your application. While the thermometers are easy to use, the remaining characteristics do not meet the ideal criteria. The accuracy of the thermometer is not within the 0.1 degree tolerance, the metal tends to degrade and rust when subjected to moisture, the life expectancy is only six months to one year, and the cost of a base model is nearly twice the budgeted price. Data for this style of thermometer were found at www.coleparmer.com, www.omega.com/thermocouples.html, and www.efunda.com/DesignStandards/sensors/thermocouples/thmcple_intro.cfm. A sample picture of this style of thermometer can be found in the Appendix.

Digital

The digital thermometer researched was found to be the best choice for your application. The thermometer was found to be easy to use, safe for use in the freezer or refrigerator, extremely rugged, accurate within 0.01 degrees, priced at $70, and expected to last more than five years. Because the unit is completely self-contained, there is no degradation of the unit and it is very easy to install and replace if necessary. As an additional feature, being digital allows for advanced interface to the thermometer unit such as datalogging for historical purposes. Again, our resources were www.coleparmer.com, www.omega.com, and www.digital-thermometers.globalspec.com. An example of this thermometer can be found in the Appendix.

Conclusion

After reviewing the four types of thermometers, glass, infrared, bimetal, and digital, we have concluded that the digital thermometer is the best choice for your application. Not only does the digital thermometer meet all of our ideal criteria, it also has added features that will prove of use for record keeping and integration into current systems. After reviewing this report, if you have any additional questions, please contact me directly or anyone in the Engineering Department.

Figure 5.4 Student Report, Comparison (with recommendation) *(continued)*

Appendix

Example of Each Type of Thermometer

Infrared Thermometer

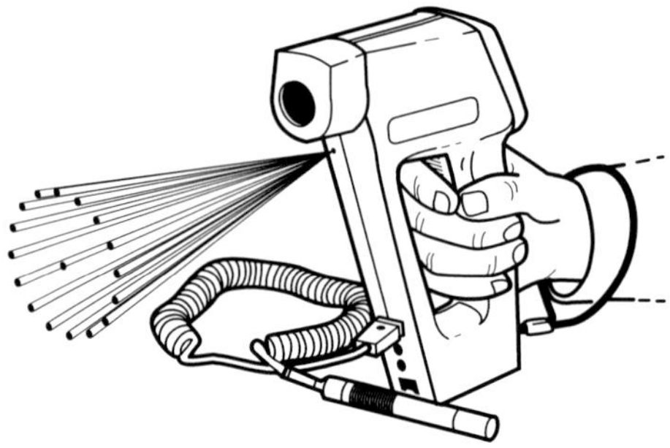

Glass Thermometer

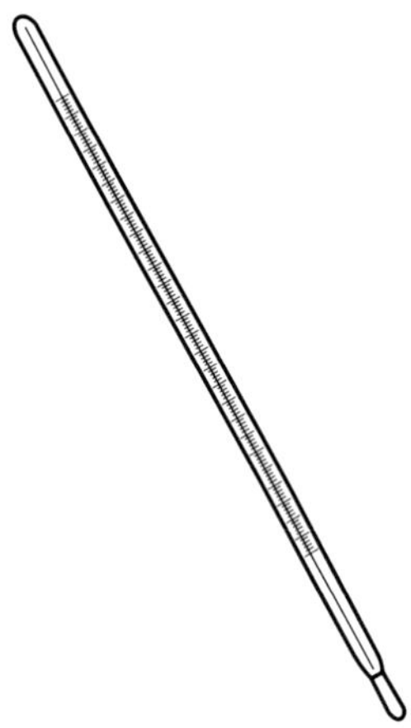

Bimetal Thermometer

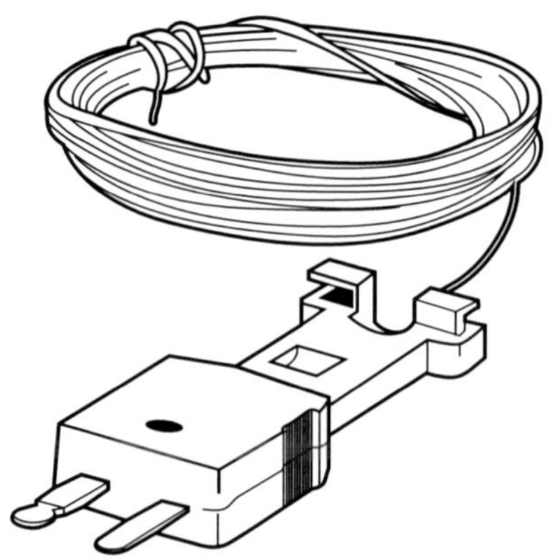

Digital Thermometer

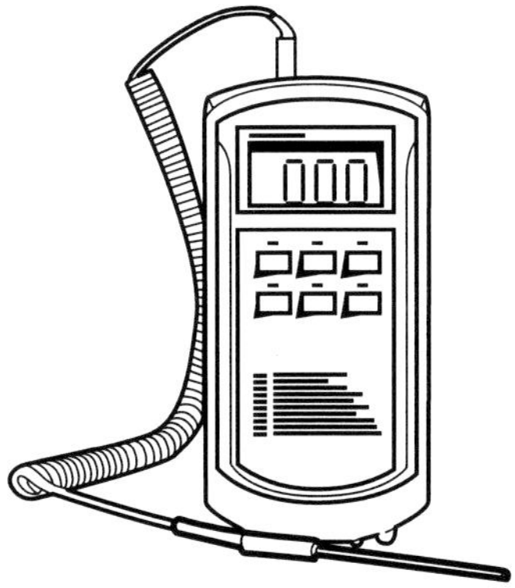

Ideal Criteria Table for Compared Thermometers

Ideal (criteria)	Real (evidence)				Judgment (difference)			
	Glass	Infrared	Bimetal	Digital	Glass	Infrared	Bimetal	Digital
Build characteristics—safe for lab use, easy to handle, rugged	can't use glass in lab, not rugged	yes—all	metal rusts	yes—all	–	+	–	+
Accuracy—within 0.1 degrees	no	yes	no	yes	–	+	–	+
Expense—$60-100	cheap—$30	no—$150	no—$120	yes—$70	+	–	–	+
Life expectancy—1-2 years	2+ years	2+ years	6 mo–1 year	2+ 1 year	+	+	–	+

Figure 5.4

Student Report, Comparison (with recommendation) *(continued)*

Figure 5.5
Sample Student Report, Comparison

Date: September 23, 2011
To: Charles Sparling
From: Paul Bailey
Subject: Comparison of Rental Services for Boom Lifts

Purpose

For the upcoming job to install an informational sign, I have put together this report comparing three different rental services for the renting of a boom lift for the install. The three rental services are Sunbelt Rentals, CAT Rentals, and In-a-Pinch Rent-It. I have used three different criteria to determine which rental service to use. The three criteria are that the price has to be under $300 a day; the lift needs to be delivered to the job site; and the lift has to reach between 50–70 feet high to complete the install. The research was done by viewing the Web sites of each rental service and finding the data that we needed to make our decision on which rental service to use.

After reviewing the information by the stated criteria, I determined that Sunbelt Rentals is the best choice for our needs. I recommend that we rent the boom lift from Sunbelt Rentals because they best fit our needs for the job.

Situation

We have an upcoming job to install an informational sign. We need a boom lift that can reach between 50 and 70 feet in the air to install. We need it to be delivered to the job site as well. We have rented lifts before, but with the changing times, we need to compare other rental services to see if we are getting the best deal possible. Although we have done previous comparisons, the need for an updated comparison is required.

The decision has been made that we needed to know about prices of lifts, whether they can be delivered, and if they can reach 50–70 feet high. I have researched them using the Internet by viewing each Web site for the services to compare them using the criteria stated. See Table 1.

Price

After deciding that money is going to be an issue during these slow times, we need a lift that is no more than $300. Two of the three services that we compared have prices that are within our criteria. CAT Rentals is the only one of the three services that is over our maximum price.

Delivery

Since we have no way of transporting a lift that big, we need the rental service to be able to deliver the lift to the job site. After reviewing the information that was gathered on the three rental services, we discovered that In-a-Pinch Rent-It service did not deliver any equipment. Sunbelt and CAT Rentals will deliver the boom lift to our jobsite with no extra cost.

Figure 5.5 Sample Student Report, Comparison *(continued)*

Lift Height

With the job that is coming up, we know that we have to have a lift that will reach 50–70 feet high in order to install the informational sign. Again, Sunbelt Rentals has the tools that we need. Also CAT Rentals has a lift that will reach to the height that we need. In-a-Pinch Rent-It Rentals only carries a lift that will reach up to 40 feet.

Table 1 Comparison Matrix of Rental Companies for Boom Lifts

Ideal (My Criteria)	Real (Actual Evidence)			Judgment		
	A	B	C	A	B	C
Price no more than $300 a day	$230	$255	$325	Yes	Yes	No
Needs to reach 50 to 70 feet	Yes	Yes	No	Yes	Yes	No
Needs to be delivered to the job site	Yes	No	Yes	Yes	No	Yes

A = Sunbelt Rentals; B = In-a-Pinch Rent-it Rentals; C = CAT Rentals

Conclusion

So, we know that we need a boom lift to install an informational sign between 50 and 70 feet high. After careful review of the criteria stated, Sunbelt Rentals is the best choice out of the three rental services, as shown in Table 1. Sunbelt will deliver the lift to the job site for us. Their price is below our maximum, which is a plus. Also, they have a lift that will reach to 70 feet, which is more than enough. Therefore, I recommend that we go with Sunbelt Rentals for our rental needs.

Thank you for taking the time to review this report and my findings. If you have any questions, please contact me.

Figure 5.6 Partial Sample Student Report, Comparison (Clint McCollum)

Note use of good criteria definitions.

Partial Sample Student Report from (Clint McCollum)

...speed, RAM size, hard disk speed and size, port compatibility, DVD Rom features, display size and cost.

Processor Speed

Based on the minimum requirements of the Adobe CS4 software, processor speed is a key factor. Adobe requires that any computer running this software have a processor speed of at least 2.0 Ghz. The HP 8730 T9400 (2.3 Ghz) and the HP 6930p T860 (2.0 Ghz) both meet the requirement. The HP Compaq 2710p is a duel core processor of 1.33 Ghz. Adobe does not consider this as meeting the system requirement.

RAM Size

Another requirement of the software is a minimum RAM size of 2 GB. All three models that I compared meet this requirement, but the default size of each model is different. The HP 8730 T9400 is installed with the most RAM with 4 GB. The HP 6930p T860 offers the least amount of RAM at just 2 GB, while the Compaq 2710p falls in between with a default of 3 GB of RAM. The latter two models would require additional memory purchased to meet what the HP 8730 T9400 offers at no additional cost.

Hard Disk Speed and Size

The hard disk size is not specified by the software requirements, but it is always a good idea to get the most memory you can get. The larger, the better. But one aspect of the hard disk that is specified is the speed. A minimum speed of 7200 RPM is needed. Of the three models we are comparing, only the HP 8730 T9400 offers both a large hard disk size (250 GB) and a hard disk speed of 7200 RPM. The other two models only offer a hard disk with a speed of 5400 RPM.

Port Compatibility

All three models offer USB ports, but only the HP 8730 T9400 and the Compaq 2710p include the IEEE1394 port. This specific port is needed to hookup and download content from the HD video camera that is used to film company events and training videos. The output from the video camera only has this type of port compatibility for hooking up to a computer.

DVD Rom Features

One of the responsibilities I have in my job is to burn video content to DVD so the Head End can upload it to the system for viewing on the Local On Demand channels. A DVD burner that burns disks that are DVD+R is required by the Head End. Each three of these models do burn to the DVD+R format.

Figure 5.6 Partial Sample Student Report, Comparison (Clint McCollum) *(continued)*

Display Size

Since the Adobe CS4 software is media intensive and includes such products as Premiere, Photoshop and Illustrator, a high quality monitor display is required. The minimum specifications are a display of 1280 × 900. The Compaq 2710p only offers a display size of 1280 × 800. The HP 6930p T860 offers a larger display of 1440 × 900, which meets the criteria. However the HP 8730 T9400 offers the largest display at a size of 1920 × 1200. The larger this display size, the more work area and canvas area I would have available to work with.

Cost

In these tough times in our industry, cost is always a factor when determining a new purchase. Each of the three models researched were vastly different in their base price. The HP 8730 T9400 is the most expensive laptop at $2500, but no additional hardware upgrades are needed. The next highest model is the HP 6930p T860 at a quoted price of $1560. The HP 6930p T860 will require an upgrade in the hard disk to meet the minimum speed required by the software. The cheapest model, at just $800, is the Compaq 2710p, but it does not have the option to upgrade the hard disk to a speed of 7200 RPM.

Figure 5.7
Student Report, Problem Analysis

To: Mary Ann Smith, SIDATS Project Manager
From: Morris Schefcik, Data Conversion Manager
Subject: Landbase Pilot Data Problems
Date: October 23, 2006

Summary

I finished my review of the landbase pilot data supplied by our subcontractor DYdata. The delivery is marred by schedule delays and poor quality. I found the key causes to be misunderstandings in data requirements and data specifications, compounded by poor communication. Our struggle with this pilot delivery not only threatens my full production schedule, but also the schedules of concurrent efforts on the SIDATS Project dependent on high-quality, timely data. Additionally, quality issues and schedule delays have the potential to jeopardize key payment milestones on the project and undermine client confidence.

This report highlights findings that came out of both our quality testing of the data and information gleaned from meetings I held with DYdata. My comments focus on three areas: late delivery, data quality, and communication.

Discussion

First, I researched all data delivery correspondences, QC results, and communications that pertained to the landbase pilot delivery. Then I scheduled two all-day meetings with the DYdata project and production managers. These meetings focused on why the delivery was late, why there was poor data quality, and why we experienced a lack of communication during production. The following summarizes the problems we experienced with the data and the understandings that came out of the meetings with DYdata.

Late Delivery

The late delivery shifted my schedule to the right by over two weeks, leaving our programmers and QC team poorly scheduled and underutilized. In addition, other SIDATS Project schedules, in particular the team developing the land navigation tool, were negatively affected by the delay due to dependence on the data.

During my meeting with DYdata, we reviewed the methods being used to communicate and track production. We found instances of erroneous production estimates and failed communication within DYdata's production teams. We concluded the following:

Figure 5.7
Student Report, Problem Analysis *(continued)*

- DYdata's e-mail notification of a delivery delay is unacceptable when received the day of a scheduled delivery.
- DYdata's estimation of production efforts is poor.
- DYdata's weekly production figures are inadequate.
- DYdata's failure to give us early notice of impending delivery delays is unsatisfactory.

Data Quality

Both data formatting and content had issues. Data files came to us formatted incorrectly requiring our programmers to "massage" data during the load process. This additional work added two days to the loading process.

The QC results reveal object based quality is 85% accurate and street centerline connectivity is 92% connected. These results are based on a 100% QC of all delivered objects and are well below the contractual requirement of 98% accuracy. We discovered significant errors during our QC:

- *Numerous errors in building names and building footprints*. DYdata failed to capture these objects from the Digital Orthophoto Quarter Quadrangle maps for areas outside corporate (city) boundaries. Furthermore, DYdata failed to capture the majority of building names for buildings located within corporate boundaries.
- *Poorly managed annotation of street centerlines*. The result was text-on-text occurrences and illegible street names.
- *Poor stitching of disparate data sets*. Many objects were left "hanging" along source edges and thus unconnected. This is the main reason for poor street centerline connectivity.

During my meeting with DYdata, we went through all data errors, reviewing related specifications and source data as we proceeded. By the end of the review session, we identified:

- Poor source material that made data work more difficult to perform
- Instances of vaguely written specifications that did not clearly state the requirements
- Misunderstanding by DYdata's programmers of the data delivery format requirements

Figure 5.7
Student Report, Problem Analysis *(continued)*

Communication

As discussed in the late delivery segment, DYdata failed to inform me of delays in production and delivery. This created resource management issues for our teams.

Additional communication issues relate to DYdata's failure to communicate during the production effort itself. Failure to ask questions and clarify ambiguities resulted in substantial errors on its part. In fact, only 10 inquiries were posted to our production Web site during the entire production effort, yet the data contained many assumptions that resulted in numerous errors.

During my meeting with DYdata, we discussed the need to communicate legitimate production estimates and all production questions that surface during production. We determined:

- There were lapses in communication by the DYdata production and management teams.
- There were assumptions made during the course of production that should have been clarified through inquiries to the production Web site.
- There was evidence of reluctance to ask questions on the part of DYdata's production managers.

In addition, we need to keep in mind other concerns. Further delays and additional problems with quality in our landbased conversion carry with it the potential for serious issues throughout the project. These include:

- Jeopardizing future data delivery milestones with the client
- Eroding our margin on the project through poor utilization staff
- Delaying client billing and revenue

Conclusion

As noted at the onset, the delivery was marred by schedule delays and poor quality. I found the key causes to be misunderstandings in data requirements and data specifications, compounded by poor communication. These schedule delays and poor quality can have far-reaching effects on the entire SIDATS Project. Please consider raising these issues when you meet with the DYdata executive board in December. Bringing these issues to the forefront and addressing them now is crucial to mitigating risk when full landbase production begins. I believe we can make up for lost ground and successfully complete the landbase conversion for the SIDATS Project if we can effectively address these issues now.

Memorandum

Figure 5.8
Sample Student Report, Problem Analysis

Note good use of headings and subheadings.

Date: November 2, 2011
To: Jeffrey Edge, Business Administrator
From: David Szpunar, Network and Systems Manager
Subject: IT Budget Overrun Analysis

Executive Summary

The IT expenditure for this year is currently $5,000 over the allotted year-to-date budget amount and is projected to reach $6,000 over by the end of the year. This problem not only affects the IT Department directly along with the senior staff members who oversee the budget, but left unchecked, it will eventually affect end-users. An in-depth analysis revealed that the primary culprits causing the problem include new recurring services and software maintenance charges, additional requirements of supporting the new building addition, and the age of the majority of existing equipment.

Background and History

The IT Department is over budget by several thousand dollars. This memo examines possible reasons why this is the case.

The yearly IT Department budget is $20,000 for expenses, and as of November 1 the year-do-date expenditures in this budget equal $25,000, and are expected to reach $26,000 by year-end. The IT budget was reduced in each of the last five years, yet has managed to come in at or under budget in each of those years. This is a good record of accomplishment, and thus the problem remains to be addressed: Why is the budget suddenly being overrun this year?

This problem is serious and either directly or indirectly affects anyone involved with the IT Department or finances, including the Business Administrator, who supervises the IT Department and oversees its budget, and the Executive Team and governing board members who provide oversight and budget approval. The cost of this problem is projected above at $6,000 this year, but the scope of the problem next year and in the more distant future is unknown but likely to grow if unchecked.

Current Effects

In the short term, all IT spending has been limited to emergency operational items only. Long term, if some adjustment is not made, the same situation will likely reoccur next year, and end-users may be affected because their equipment may not be maintained or replaced as needed.

Figure 5.8
Sample Student Report, Problem Analysis *(continued)*

Analysis

In order to determine the cause of the budget problems, I reviewed detailed logs of transactions that have taken place in the budget for this year. The trends from that review, along with my personal experience in the IT Department over the past several years, lead to the following three main possibilities. Any number of them may be contributing factors to the problem.

1. *New building addition equals added support costs.*

The major building addition completed earlier this year began with its own construction budget from which all the new networking and computer equipment was purchased that was designated for use in the new building. This included new switches and routers, battery backups for this equipment, wiring of most areas in the new building with network jacks, computers for the Youth Internet Café area, new laptops for the youth staff, and other miscellaneous items.

However, now that the building is complete and the construction budgets are spent and closed, the support and maintenance of the new building equipment falls under the existing IT budget, which, as mentioned previously, has been level or declining from year to year. Additionally, supporting the cost of maintenance to more equipment is not working with the same or less budget.

2. *Existing equipment is aging and needs to be repaired/replaced more often.*

In addition to the increase in new equipment to support, at the other end of the spectrum is the equipment we already owned prior to the new construction. Because the IT budget was reduced in prior years, IT equipment and computers have been used past their useful life in several instances. The older computers are beginning to fail at inopportune times, preventing users from accomplishing many of their job functions while a replacement is procured.

These unscheduled replacements are taking large amounts of budget money. Equipment failure is costly because it pulls IT resources from other areas, and it requires purchasing individual computers to replace the old equipment (while discounts are usually available for purchases of multiple computers). Research into the transaction history has shown that seven computer workstations have needed unexpected replacement this year. In addition, twenty miscellaneous computer accessories have needed replacement due to failure in the same timeframe (speakers, soundboards, keyboards, mice, etc.).

3. *New subscriptions are eating away at the monthly budget.*

Several services are charged to the IT department on a recurring basis to provide ongoing support or features to the staff, directly or indirectly.

Figure 5.8 Sample Student Report, Problem Analysis *(continued)*

These services include our long-standing subscription to uptime monitoring (and alerting of downtime) of internal office servers and external Web servers, for which we spend approximately $20 per month. Newer subscriptions include a remote-control subscription to allow administration of several specific computers from remote Internet connections, which costs another $20 per month. The service that we use to filter our incoming email for spam and viruses is approximately $90 per month. These subscriptions alone are costing about $1560 per year, and all were added to the budget within the last one to two years.

We also have ongoing software maintenance agreements and subscriptions that add a large amount to this number. The cost of our antivirus software and updates is about $1600 per year. Various other software products, which we pay for either every year or every two years, are costing an additional several thousand dollars to maintain, even at the non-profit rates we enjoy from most vendors. Records of the exact amounts and renewal rates for each contract have not been kept and need to be compiled to manage this area of the budget properly, and thus only an estimate can be provided here. Over time, the additional subscriptions have started to consume more of the available budget, leaving less for the general operational costs that used to comprise a higher percentage of the budget.

Conclusion

The IT budget has overrun by $5,000 year-to-date with a projected $6,000 overrun by year's end. The primary three reasons provided herein for these IT budget overruns (new building support, failing older equipment, and increased subscriptions cost) need to be addressed before the end of this year in order to prevent as much of the budget overrun as possible and make sure any corrections are made to prevent a repeat of the situation next year. There may be other reasons that become obvious from further analysis, but I believe this report provides a good foundational analysis of the problem. I would be happy to work with anyone necessary to dig deeper into the problems, collaborate on possible solutions, and assist with the implementation of the preferred solution(s) once a decision has been made.

Figure 5.9
Student Report, Recommendation

Sky Company, Inc.

To: Mr. Clive Smith, President
From: Jim Wilson, Research Associate
Date: June 4, 2007
Re: Online Writing Lab Research and Recommendation

Summary

As you instructed, I evaluated several Online Writing Labs (OWLs) for use by Sky staff. My research focused on information and features in the OWLs that will help our employees improve the quality of website recommendations made to clients. Below are the four OWLs approved for consideration:

1. http://www.library.cornell.edu/okuref/research/webeval.html
2. http://owl.english.purdue.edu/handouts/research/r_evalsource.html
3. http://www.library.ucla.edu/libraries/college/help/critical/index.htm
4. http://www.library.vcu.edu/help/evaluate.html

Each site was scored based on eight criteria listed in order of importance.

- Tips/questions to consider when evaluating Web site content
- Additional resources provided
- Easy navigation
- Good visual design
- Recent sources cited
- Traffic ranking
- Inception
- Number of sites linking to the OWL

To rank the sites, I visited each and counted the relevant tips and questions offered for Web site evaluation. I also recorded the number of additional resources given and noted whether recent sources were referenced. Both navigation and design scores were subjective impressions from my visit. None of the four sites excelled in navigation or design, in my opinion, but all were adequate. Traffic ranking, inception date, and number of sites linking to each OWL came from http://www.alexa.com/.

The best scoring sites are maintained by Cornell University (site number 1) and Purdue University (site number 2). The main page of the site at Cornell contains several lists of important factors to consider when looking at a Web site. Each list group has a link that takes you to a new site with detailed tips and questions.

Figure 5.9 Student Report, Recommendation *(continued)*

The OWL hosted by Purdue also offers a good list of questions to ask when looking at a Web site. This site exceeds expectations by offering a plethora of additional resources, including general search tips, metasearch engine help, and evaluation tutorials. In the discussion section of the report I go into more detail explaining why I believe Purdue University offers the best OWL for Sky Company to use.

Discussion

Background

Our challenge to recommend a few great Web sites for client use grows harder every year. How do we find the best sites when competitors such as Yahoo! and Google boast the ability to search billions of Web pages? The answer lies in our ability to recognize quality Web sites and quickly eliminate those sites that fail to meet our rigorous standards. We must do what Yahoo! and Google cannot—evaluate the credibility, reliability and relevance of the Web material before it is passed on to the client. Recently, Sky Company's ability to perform this task has come into question by several important clients. To improve results and differentiate our services from competitors, we need a method for site evaluation available to all Sky researchers. Incorporating the regular use of an Online Writing Lab that provides trustworthy resources and thoughtful insight is important for Sky Company's continued success and future excellence.

Criteria

In the Research section that follows, I discuss the first five criteria for each of the four OWLs being considered. The last three criteria results were obtained from http://www.alexa.com/ and are included in the appendix.

Listed in order of importance for evaluating each site:

- Tips/questions to consider when evaluating Web site content
- Additional resources provided
- Easy navigation
- Good visual design
- Recent sources cited
- Traffic ranking—included as a measure of popularity among all users
- Inception—established history of service availability
- Number of sites linking to the OWL—measure of popularity from sites with similar interests

Figure 5.9
Student Report, Recommendation
(continued)

Research

http://www.library.cornell.edu/okuref/research/webeval.htm

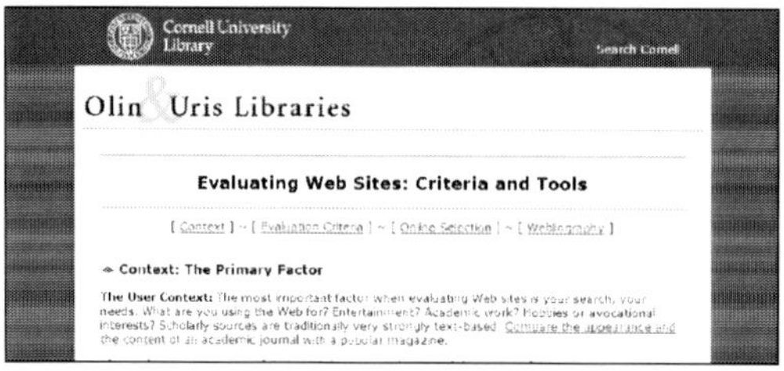

Traffic, 1,985

Inception, 7/15/85

Linking sites, 3,470

The site from Cornell University gives five different sets of criteria for judging Web sites. Each has a link to a detailed list of items or questions that should be used to evaluate a site. Many points are essentially repeated between the sets; however, the repetition helped to reinforce the most important considerations. The sources cited were appropriate and recent.

An unexpected bonus feature appears near the bottom of the webpage under the heading, "Online Selection Examples: Subject Portals." The portals divide content by subject and offer another method for Sky employees to find material. One disappointment is that the links open within the same browser instance, taking you out of the Cornell site. Bookmarking or tabbed browsing is needed to overcome this shortcoming.

http://owl.english.purdue.edu/handouts/research/r_evalsource.html

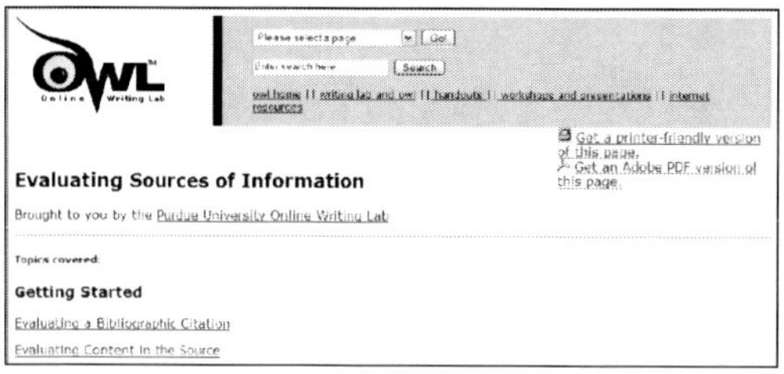

Traffic, 3,232

Inception, 4/24/85

Linking sites, 4,939

The OWL hosted by Purdue University divides the section "Evaluating Sources of Information" into five pages: Getting Started, Evaluating a Bibliographic Citation, Evaluating Content in the Source, Evaluating Internet Sources, and Further Resources. The "Evaluating Internet Sources" page offers a single set of questions particularly suited for Web sites. All five pages link to each other, creating a unified strategy for evaluating information found on the Web. Directing employees to use this site would establish a standard for evaluating Web sites.

As I mentioned in the summary, the Purdue OWL has several other helpful resources. Our employees could benefit from working through the tutorials or trying a new metasearch engine. The site lists many external sources for research and reference, as well as links to other OWLs. I also found this site to have the best navigation and most recent sources cited. The site does not address cost; however, a fair use policy is provided at http://owl.english.purdue.edu/owl/resource/551/01 and it covers use of the site for commercial purposes.

Figure 5.9 Student Report, Recommendation *(continued)*

http://www.library.ucla.edu/libraries/college/help/critical/index.htm

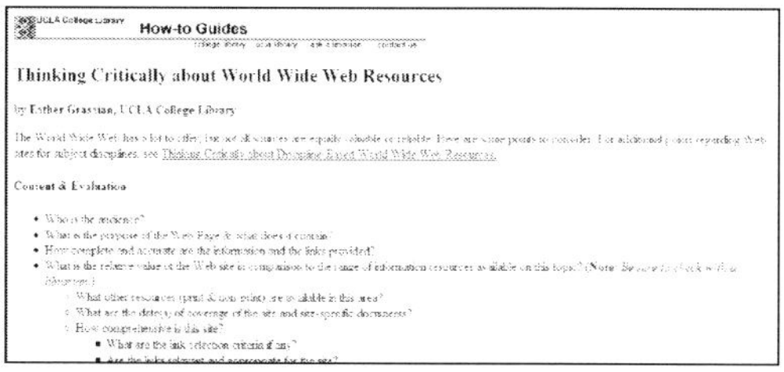

Traffic, 3,232

Inception, 4/24/85

Linking sites, 4,939

The page connected to the hyperlink above does list good questions to ask about Web sites, but it offers little else. Links at the top of the page for the college library and UCLA library have different navigation structures and are not connected. Both sites have information that could be useful; however, it is not organized in an intuitive manner. One resource link that caught my attention in the UCLA library Web site is Google Scholar™. I believe we should consider adding the Google Scholar beta to our repertoire of search engines.

http://www.library.vcu.edu/help/evaluate.html

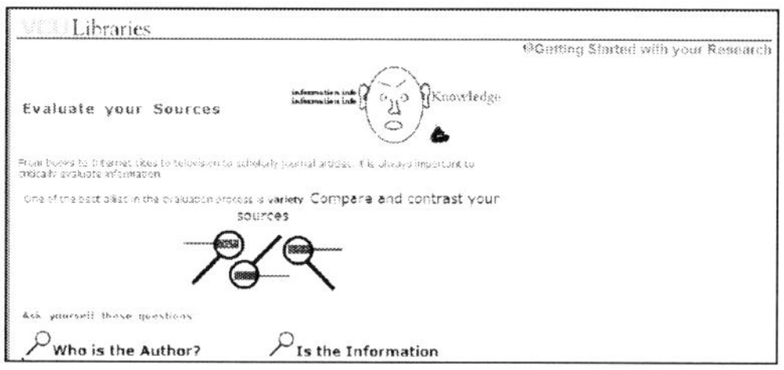

Traffic, 11,454

Inception, 11/11/88

Linking sites, 692

I almost missed the navigation for the VCU site. A less-than-obvious link at the top and bottom of the page is the only way to get to the main VCU navigation. Just as bad, most of the links in the VCU OWL take you to external sites without launching a new browser instance. I found it impossible to research this way or give this site serious consideration for Sky employees to use.

Recommendation

Of the four sites investigated, I found the VCU and UCLA OWLs least appropriate for use by Sky Company. Although some helpful information is included in both, difficulty navigating through the sites makes them poor choices. The OWL created by Cornell University also creates navigation issues by linking away from the main site, but I still found it moderately easy to follow. Having all the evaluation tips on one page was a big plus for the Cornell site. By far the most impressive OWL is hosted by Purdue University. The Purdue site met all criteria, has consistent navigation, and offers additional features our staff could use.

One item that still needs to be addressed is to obtain permission to use the OWL at Purdue for commercial purposes. None of the sites approved for review offered an unconditional commercial option or pricing for business use. Please let me know if I should contact the site administrator about the process to obtain permission or for any additional information you would like.

Appendix

Three-Column OWL Comparison

A = *http://www.library.cornell.edu/okuref/research/webeval.html* Cornell University
B = *http://owl.english.purdue.edu/handouts/research/r_evalsource.html* Purdue University
C = *http://www.library.ucla.edu/libraries/college/help/critical/index.htm* UCLA College
D = *http://www.library.vcu.edu/help/evaluate.html* Virginia Commonwealth University

Ideal (criteria)		**Real (evidence)**			**Judgment (difference)**			
Measures of Performance	**A**	**B**	**C**	**D**	**A**	**B**	**C**	**D**
List of items/questions to ask to evaluate Web site content (> 20 items)	20+	30+	30+	17	+	+	+	–/+
Additional resources provided (at least 5)	14	10	1	21	+	+	–	+
Easy navigation	ok	yes	no	no	–/+	+	–	–
Good visual design	ok	ok	ok	ok	–/+	–/+	–/+	–/+
Recent sources cited (within 3 years)	yes	yes	no	yes	+	+	–	–/+
Traffic ranking from Alexa (daily reach > 2000)	1,985 (no reviews)	3,232 5 stars	2,343 (no reviews)	11,454 (no reviews)	–/+	+	+	+
Online since (before 1996)	7/15/85	4/24/85	4/24/85	11/11/88	+	+	+	+
Other sites linking to (1000+)	3,470	4,939	2,090	692	+	+	+	–

Overall Claim: Based on these data, http://owl.english.purdue.edu/handouts/research/r_evalsource.html produced by Purdue University is the best resource to help Sky Company employees evaluate Web sites. Performance items are listed in order of importance.

Figure 5.9
Student Report, Recommendation *(continued)*

Figure 5.10
Sample Student Report, Recommendation

Note good use of persuasive language.

To: Jim Harris, Manager of Operations
From: Chris Martin, Manufacturing Safety Auditor
Date: November 2, 2011
Subject: Safety Upgrades

Introduction

In response to our meeting for improved safety equipment, I am recommending a new model of safety glasses to replace those currently used. As we discussed in our last meeting, the number of near miss incidents has increased since implementing new line processes in the manufacturing department. We agreed this situation has resulted in a sense of urgency to reevaluate and prioritize the issue of employee safety. I have included for your consideration a description of the safety glasses I am recommending, reasons these glasses will offer significant improvements, results of an employee test survey, and information regarding OSHA compliance.

Discussion

As we discussed, the increase in near miss incidents has been associated with the recent equipment upgrades in the manufacturing department. When this change was accomplished, a hazard assessment for the new equipment did not accurately take into consideration the differences in operation of the new machines and how ineffective the currently used Personal Protective Equipment would be. Ninety-eight percent (98%) of all near misses reported were related to the eye area while safety glasses were being worn. Employees reported poor fitting glasses, inadequate side shields, and the glasses sliding down their nose or nearly falling off when looking down at the control pads on the process machines. As you are aware, the use of safety glasses is required in this area because of the probability of loose material discharging from the machines during operation. The new machines move much more quickly, increasing the likelihood an employee may become injured if preoccupied with adjusting their safety glasses. With this danger in mind, I have addressed the benefits based on fit and employee feedback.

Fit

To resolve the issue of fit, it was important to understand the differences between the glasses we are using and those I recommend.

- The new glasses have a wraparound design with tapered ear pieces that do not require adjustment and stay in place securely while being worn. Our currently used glasses are not adjustable, and the frame is rather large and does not accommodate all face sizes, which allows for gaps around the eye area.

Figure 5.10
Sample Student Report, Recommendation *(continued)*

- The area around the eyes is much better protected due to the contour of the design. The wraparound aspect is even more significant because side shields are not necessary. The side shields on our current glasses allow for gaps at the top and corner of the eyes and have been reported to obstruct peripheral vision.
- In addition to these features, the ends of the earpieces allow for a safety lanyard to prevent losing the glasses.
- Those employees who participated in the test group reported high satisfaction in comfort and an increase in confidence with the level of protection offered.

Employee Evaluation Results

I was able to obtain samples of the new safety glasses and distributed them to a group of 30 employees to evaluate. Here is a compilation of those results based on comfort, effectiveness, and confidence:

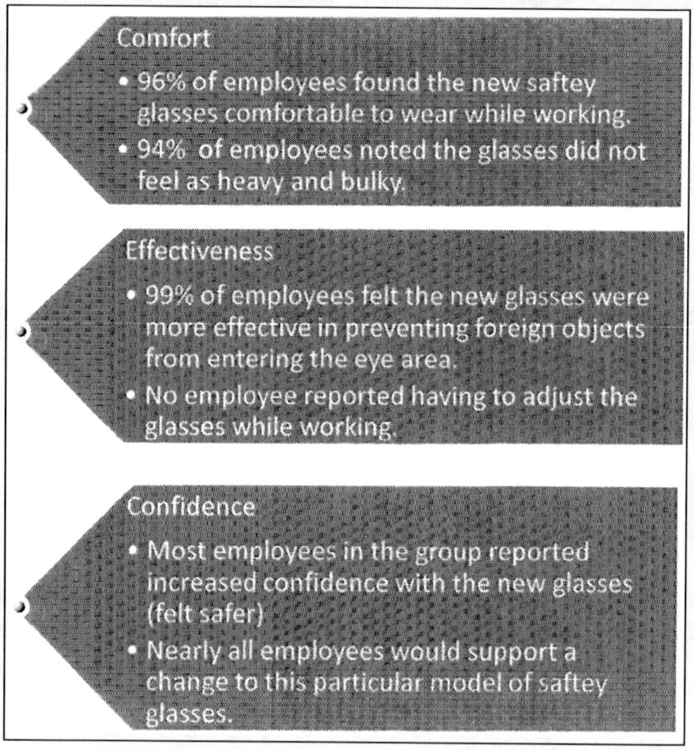

Figure 5.10 Sample Student Report, Recommendation *(continued)*

Compliance

The importance of employee safety cannot be emphasized enough. There have been many incidents of employee injuries where the employer was found at fault for violating OSHA requirements. One of the most important aspects of employer responsibilities is to provide employees with proper Personal Protective Equipment (PPE) and training to ensure they understand proper usage. Failure to do so has resulted in OSHA violations from random inspections and inquiries based on employee reporting. These fines can be a serious financial burden in comparison to the minimal cost associated with a simple upgrade of equipment. OSHA violations start at $7,000 for the first citation and will increase to $70,000 if not corrected within six months.[1]

Transition

An upgrade to this particular model of safety glasses is relatively inexpensive in comparison. The initial cost is estimated at an additional one dollar per pair over the currently supplied safety glasses. A change to this new model will not cause an excessive increase in expenditures and can be ordered for the next quarter. Since we are retaining our current vendor, we will only need to change the model number on the order form. This is a very feasible and affordable venture that will serve in the best interests of our employees' safety.

Recommendation

Given the nature of the current situation involving the safety of our employees, I recommend we act soon to upgrade our Personal Protective Equipment. The near miss reports, submitted by our employees, have identified a potentially life-altering hazard that we must take into consideration with the utmost urgency. I strongly believe an upgrade to the model of safety glasses I have recommended will satisfy this deficiency in safety. I hope I have satisfied your query regarding our previous discussion. If you have any questions or would like more information on the OSHA standards or the safety glasses, please contact me as soon as possible.

[1]Code of Federal Regulations: Part 29, General Industry Standards. More information is available at www.osha.gov.

Figure 5.11
Partial Sample Sample Report, Recommendation (Eden Dorsett)

Note good use of comparative/contrastive language and emphasis.

Partial Sample Student Report (Eden Dorsett)

Recommendation

Given the three compact utility tractors (within the 45 HP range) I evaluated, it appears that the Kubota L4240 is the least suitable tractor for our business. Although the Kubota L4240 has the highest lift capacity, it does not come with a locking differential or automatic 4-wheel drive engagement, its hood is not sloped for optimal visibility, and it is the highest priced tractor evaluated.

The second tractor I evaluated, the John Deere 4120, is cheaper than the Kubota L4240. It does have an option of a locking differential, and comes with a sloped hood for greater visibility; however, the John Deere 4120 does not have an option for automatic 4-wheel drive engagement. Therefore, I believe the John Deere earned the second place ranking.

Based on the evaluated criteria, I believe the New Holland (Boomer 3045) is the most impressive tractor I compared. The New Holland (Boomer 3045) earned a positive score on every criterion except for the lift capacity, where it came in last. Even though the New Holland tractor has the lowest lift capacity, I am not swayed on my decision because each of our rear-mounted 3-PT hitch implements weighs much less than 2,000 lbs. The New Holland is definitely the cheapest tractor at $21,992. Along with being the cheapest of the three tractors, it also comes standard with a locking rear differential, sloped front hood for excellent visibility, automatic 4-wheel drive engagement, standard 2-year warranty, option of shuttle-shift/hydrostatic transmission, and a mid-mount PTO. The mid-mounted PTO on the New Holland tractor will enable us to power a front mounted snow blower; this will work well in our snow removal service. Based on the three tractors I evaluated, I strongly recommend you purchase the New Holland (Boomer 3045) tractor.

Benefits and Drawbacks

Everyone in our company will benefit by our company purchasing this New Holland (Boomer 3045). As you know, it is inevitable that we purchase a new utility tractor because our old Massey-Ferguson could break down at any moment. Knowing that a new tractor purchase is necessary, our company will benefit by purchasing the New Holland tractor because it is the cheapest of the three compared. Therefore, we are spending as little of the company's money as possible. Also, by having a tractor that will keep the business up and running, we will have this expenditure paid for in short order. This benefit will even cascade down to the department and individual level.

This new tractor will be much more efficient than our current one. With this in mind, our grounds crew department will not have to hire more

Figure 5.11
Partial Sample Sample Report, Recommendation (Eden Dorsett) *(continued)*

manual labor, due to the extra functionality of the New Holland. By keeping the business up and running (without missing a beat), we will not have to consider laying anyone off. The bottom line is this: By purchasing this New Holland tractor, our normal income will be sustained, thereby enabling us to continue paying the employees.

There is one possible drawback to making this purchase, and that is the learning curve everyone will be up against when first learning to operate this tractor. In all fairness, this learning curve would also be present with the other two tractors I evaluated as well. Therefore, to mitigate this learning curve, I plan on setting aside half of a workday for training, which should clear up a lot of questions the other employees may have. Also, everyone in training will obtain some hands-on experience with our new tractor; this will hopefully allow us to "hit the ground running."

Timeframe and Cost

Time is critical in making this purchase. The local New Holland dealership has a Boomer 3045 on its lot. If we do not purchase this one, we will have to wait until the dealer can get another one in stock, which can take from one week up to a month. In regards to our current Massey-Ferguson tractor, we are working on borrowed time because it could break down at any moment. Therefore, in my estimation, we should act as soon as practical.

As I stated earlier in this report, the New Holland (Boomer 3045) is the cheapest of the three tractors I compared. In many cases, the old adage "You get what you pay for" is somewhat true; however, in this case we are getting a very high quality tractor and still paying less than we would for the other two tractors. In fact, I believe we are getting quite a bargain. Before taxes, the total cost of the New Holland (Boomer 3045) is going to be $21,992. In all honesty, all three tractor manufactures—Kubota, John Deere, and New Holland—are three of the leading tractor makers in the world. Therefore, I have total confidence that this tractor will be very reliable for years to come, and not to mention well worth the price.

Figure 5.12 Sample Student Report, Lab

Note: This sample is a fun approach to a lab report.

TCM MythBusters

Myth: "Tapping a Soft Drink Can Prevent Foam Overflow"

MythBuster: Russell Washer

Figure 5.12 Sample Student Report, Lab *(continued)*

Abstract

I set out to try and find if tapping a shaken can of soda would really prevent its contents from overflowing with foam. In my experiment, I shook two different brands of soda, Coca Cola and Mountain Dew, for ten seconds and then opened them: without tapping on the can, after tapping on the can's top for ten seconds and finally tapping on the can's side for ten seconds. I concluded that this myth is plausible, depending on the type of soda.

Background Information

We all at some point have opened up a soft drink that has fallen to the ground or has been shaken up. In the end, we get a nice foaming mess overflowing out of the can. To see why this occurs, I did some online research to find out what causes this phenomenon. According to http://www.stevespanglerscience.com, http://www.lifescience.com, and http://www.snopes.com, a water and carbon dioxide are forced into the can at a very high pressure and then sealed inside during factory processing. This amount of pressure fuses the carbon dioxide and liquid within the can. When opened, the pressure releases and gas escapes in the form of bubbles as they multiply, expand, and force the liquid to the top, thus releasing carbon dioxide into the air. This creates what we know as *fizz*. When a can is agitated, or shaken, the carbon dioxide separates from the liquid and forces the bubbles to the side of the can as they intensify in volume. As the volume of the bubbles increases, the more gas is being built up within the can. When opened, the pressure escapes, forcing these bubbles to push the liquid up and release the carbon dioxide at a much faster rate.

Hypothesis

If an agitated soft drink is tapped, does this prevent its contents from overflowing? According to http://www.stevespanglerscience.com, if a person was to tap on the can's top or side and then open it, this would cause the CO_2 bubbles to float above the liquid and prevent the liquid from escaping. In my hypothesis, I posited that tapping the top or side of the can would not be enough to prevent the soda from foaming over. The purpose of this experiment was to see if my hypothesis is indeed correct, or if this old myth was true.

Material

In my experiment, I included the following soft drinks:

- Three unopened Coke cans
- Three unopened Mountain Dew cans

Figure 5.12
Sample Student Report, Lab
(continued)

I decided to use two different brands to see if this process applied to just a certain brand of soda or if this occurred in all brands of sodas.

Methodology
I decided to test both brands of soft drinks through the same three experiments.

Experiment 1—Without tap
Step 1: I grabbed a can of Coca Cola.
Step 2: I shook the can for ten seconds.
Step 3: I placed my fingernail under the can's tab and pulled up.
Step 4: I observed results.

Experiment 2: Tapped on Top
Step 1: I grabbed a second can of Coca Cola.
Step 2: I shook the can for ten seconds.
Step 3: I then tapped on the can's top for ten seconds.
Step 4: I repeated step 3 in Experiment 1.
Step 5: I repeated step 4 in Experiment 1.

Experiment 3—Tapped on side
Step 1: I grabbed a third can of Coca Cola.
Step 2: I shook the can for ten seconds.
Step 3: I tapped on the can's side for ten seconds.
Step 4: I repeated step 3 in Experiment 1.
Step 5: I repeated step 4 in Experiment 1.
Step 6: I repeated all three experiments with the Mountain Dew cans.

Results
To my surprise, my hypothesis was partially correct. When I shook the cans without tapping on them, their contents overflowed. The Coca Cola overflowed in a constant stream down the side of the can for a few seconds. Tapping on the side and top of the Coca Cola can seemed to have little or no effect on the amount that overflowed. When I shook the can of Mountain Dew and opened it, however, the amount of overflow was very little as drops of foam spilled over the side. When I tapped on the can's top and side, there was only a small amount of foam that exited the can. A majority of the foam stayed contained inside. The results can be seen in Table 1. Visuals are also provided in Figures 1.1 through 1.6 in the Appendix.

Figure 5.12
Sample Student Report, Lab *(continued)*

Table 1

Action	Coca Cola	Mountain Dew
Shaken 10 seconds	Constant Overflow	Small drops flow over the can
Shaken and tapped on top 10 seconds	Constant Overflow	No Overflow
Shaken and tapped on side 10 seconds	Constant Overflow	No Overflow

My final verdict to this myth is that it is plausible, seeing how this procedure worked on one brand of soda and not the other.

Analysis

As I shook the cans, I could feel the cans start to tighten as the pressure built inside. The Coca Cola can seem to tighten way more than the Mountain Dew can. It would seem that each can contained a different amount of carbon dioxide build-up. The Coca Cola would appear to contain a greater amount of carbon dioxide than the Mountain Dew, thus causing the Coca Cola to have a greater amount of overflow. Although tapping a can be effective, it is not 100% guaranteed, as I have shown in this experiment.

Experience shows us that letting a soft drink sit for a period of time allows its contents to settle and prevents a foaming overflow. Still today, this seems to be the best remedy.

Figure 5.12 Sample Student Report, Lab *(continued)*

Appendix

Experiment Results

- Coke

Figure 1.1 Shaken for 10 seconds

Figure 1.2 Shaken and tapped on top for 10 seconds

Figure 1.3 Shaken and tapped on side for 10 seconds

Figure 5.12
Sample Student Report, Lab *(continued)*

- Mountain Dew

Figure 1.4 Shaken for 10 seconds

Figure 1.5 Shaken and tapped on top for 10 seconds

Figure 1.6 Shaken and tapped on side for 10 seconds

Side Note

I originally also included Sprite, but when I shook the can without tapping, no contents overflowed. So, I made the personal decision to drop this brand from the experiment.

Figure 5.12
Sample Student Report, Lab
(continued)

History

According to http://www.Wikipedia.org, in 1772, Joseph Priestly invented carbonated water by discovering how to combine or infuse both carbon dioxide and water. He suspended a bowl of water in a beer vat and dripped oil and sulfuric acid onto chalk to create the carbon dioxide, allowing it to settle within the agitated water.

References

- Bryner, M. (n.d.). Retrieved November 4, 2009, from http://www.livescience.com/mysteries/061010_soda_fizz.html
- Mikkelson, B. (2007, July 21). Retrieved November 3, 2009, from http://www.snopes.com/science/sodacan.asp
- Priestly, J. (n.d.). Retrieved November 3, 2009, from http://en.wikipedia.org/wiki/Joseph_Priestley
- Soda Can Shake-Up. (n.d.). Retrieved November 3, 2009, from http://www.stevespanglerscience.com/experiment/00000024

Chapter Summary

This chapter introduced you to some important concepts and considerations when producing an informal workplace report. Specifically, the chapter introduced you to a comparison report, a problem analysis and problem/solution report, a recommendation report, a progress report, and a lab report. Although they differ in purpose and execution, these reports also have many common characteristics.

If you follow the logical steps outlined in this chapter, you will master the process of planning, researching, and drafting these reports. Like anything else, comfort with the process will come with experience.

End of Chapter Activities

ACTIVITY 1 Group Comparison Practice

In small groups, discuss the following scenarios and outline how you would meet the challenge.

Scenario 1:

Your company is having a sales rollout at HQ, where you work. As a member of the event planning staff, you have been tasked with selecting the t-shirt vendor. You will make the recommendation to your boss.

Brainstorm a list of information you will need to know/consider to accomplish this task. What criteria will you establish? How will you define the "ideal" vendor?

Scenario 2:

Your department is hosting a meeting of all its employees, local and remote, in a couple of weeks. Your boss has asked you to recommend a restaurant where you can all eat dinner. She has given you a specific budget. The out of towners will be staying at the same hotel and will be sharing transportation.

Brainstorm a list of information you will need to know/consider to accomplish this task. What criteria will you establish? How will you define the "ideal" restaurant?

Scenario 3:

The upper management of your company has decided that it will fund an initiative for healthier employees. To that end, the company has agreed to help fund memberships to a health club or gymnasium. Your manager has charged you with investigating three health clubs in order to help management choose which club to use.

Brainstorm a list of information you will need to know/consider to accomplish this task. What criteria will you establish? How will you define the "ideal" vendor?

ACTIVITY 2 Group Problem Analysis/Problem Solution Practice

In small groups, discuss the following scenarios and come up with strategies you would use to meet the challenge.

Scenario 1:

You work in a retail store where you are a supervisor in one of the departments. Although historically your department has been a solid revenue producer, lately the numbers have dropped off, and the store manager has expressed some concern to you.

You want to understand the reasons so decide to do an analysis of the situation. You will report your findings to the store manager. What do you do? (You may invent data to report.)

Scenario 2:

You are a supervisor of a customer service department in a medium-sized company, with ten people reporting to you. The company has been expanding, which means the product line that you support has grown in numbers and complexity. Your boss, who has worked his way up the company ladder for several years and has a reputation of being somewhat inflexible and cheap, has steadfastly refused to let any of your people go to any training classes for the last two years.

Although you have managed this issue, it's beginning to reach a critical stage, with people having genuine training needs on new products and systems.

You need to put together a report to document what you see as the problems. How do you do that?

Scenario 3:

You work as a project manager for a small construction company. Your primary responsibilities are to make sure that the project comes in on time and within budget. This challenge involves keeping the various subcontractors coordinated and on point. Recently, you have faced a number of unforeseen problems, and you are concerned about the ramifications of those problems. Having done some research, you understand it is time to write a report to alert your manager to the issues. How do you go about that? (Invent some reasons and discuss how you would provide data for them.)

CHAPTER 6

Instructions/Manual

Overview

You will likely be asked to write a set of instructions at some point in your career. "Not a problem," you say. "I tell people all the time how to do things." Yes, you probably do, but while you're telling them, you're also using body language, nonverbals, or even props to help them understand your message. You're also there to answer any questions they may have.

When you're asked to *write* a set of instructions, it's a completely different story. Poorly written instructions can lead to miscommunication and errors or delays in projects, both internally and externally, and can lead to ill will for your company or organization. Ultimately, poorly written instructions can lead to serious injury or even death. So, it's vitally important that you be able to write effective instructions.

When planning to write instructions, analyze your users thoroughly and consider carefully the words you will use and the layout and visuals that will help the users understand your instructions. Layout and visuals are paramount to usability. The instructions should be accurate, usable, and complete.

An effective way to write a set of instructions is to record what you do yourself as you perform the task or to record what someone else does as you watch him perform the task. Or you can have someone else document what you do as you perform the task. Taking quality photos of each step as it is performed will provide you with the necessary visuals for your set of instructions.

Instructions come in all shapes and sizes and cover every topic you can imagine. Instructions can be informal or formal, complex or simple. This chapter discusses information to help you write an accurate, complete, accessible set of instructions.

Audience Analysis

In the planning stage, you'll want to analyze carefully a representative user for whom you are writing the instructions. For example, are you writing the instructions for a novice or for an experienced user? Many people have an aversion to reading manuals because so many of them are poorly written. It becomes your responsibility as the writer to sway users to want to read your instructions by the approach you take (content and design). If you have a clear picture of the user, you are far more likely to choose an appropriate approach. Chapter 2 discusses audience analysis at length, so the discussion here is limited to reminding you of the importance and necessity of doing one.

The Set of Instructions

All instructions involve front matter, middle matter, and back matter. The rhetorical context (topic, reader, purpose, and situation) determines how you approach each of these elements.

FRONT MATTER

The front matter sets the context for your reader. Here's where you motivate the reader to read the instructions and where you tell the reader how to prepare to carry out the instructions. Front matter typically includes a title page, a table of contents, an introduction, materials needed, and safety information. See Figure 6.1 for a student sample of front matter.

Title Page and Table of Contents

Have you motivated the user to read your set of instructions? An effectively formatted and attractively presented title page may be the ticket to getting users to at least take the first step to reading your instructions. Readers like to know what they're getting ready to read, so a table of contents helps make the material accessible to the user. It allows the reader to jump to the section he wants to read.

TIPS FOR CREATING AND FORMATTING A TABLE OF CONTENTS (TOC)

- Label the page "Table of Contents" or "Contents."
- Begin each major heading *flush left*.
  - If you are including subheadings, indent them like this under their primary heading.
- Word each heading and subheading *identically* to the matching heading/subheading in the manual.
- Place page numbers directly opposite the heading or subheading *flush right*.
  - Note: The word *page* is not necessary. You may use running periods (....) if you like.
- Include only the number of the first page where you begin discussing the topic. For example, if the discussion of "Tips for Creating Visuals" begins on page 11 and continues to page 13, only the "11" should appear in the TOC.

Microsoft Word can help you create your Table of Contents. Go to the "References" tab; the first section is entitled "Table of Contents." It directs you on how to create your TOC.

Figure 6.1
Front Matter from Former Student Eden Dorsett

How to Fell a Tree

By Eden Dorsett

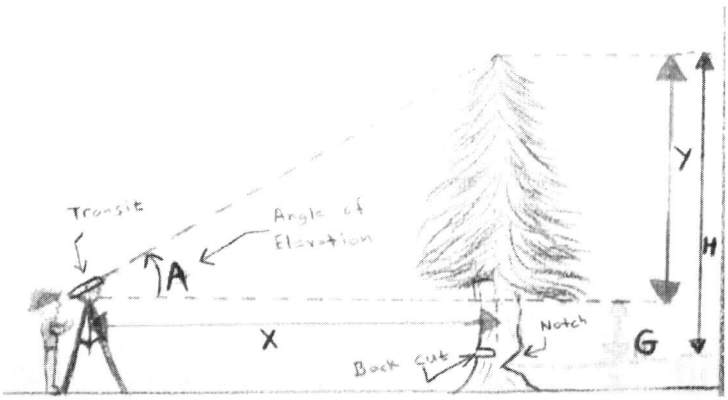

Graphical depiction by author—Eden Dorsett

Figure 6.1 Front Matter from Former Student Eden Dorsett *(continued)*

Table of Contents

Figure 6.1
Front Matter from Former Student Eden Dorsett
(continued)

Introduction

- This manual is written for individuals who have little or no experience in felling (cutting down) trees. However, an experienced user may find certain sections useful based on their individual tree felling situation. It is recommended that a beginner start with Section 1 and work his/her way through all of the sections. Please refer to the Definitions section, on page 2, for words you may not be familiar with.
- The purpose of this manual is to take you through the process of how to cut down trees to be used as firewood.
- This manual is intended to be used by adults who have a desire to fell their own trees and cut up their own firewood (for the purpose of heating a private residence). This manual is not intended to replace any formal training for commercial logging.
- It is assumed that the person using this manual has access to a chainsaw and is somewhat familiar with its operation.

Materials Needed

- Chainsaw
- Open-Ended or Boxed Wrench (To Make Chain Tension Adjustments)
- Screwdriver – Standard Tip (To Make Chain Tension Adjustments)
- Chopping Axe
- Splitting Maul
- Splitting Wedge
- Sledge Hammer
- Log Chain (Minimum Length of 10ft)
- Transit or Homemade Angle Measurement Tool (Optional)

Cautions/Warnings

Warning—Tree felling can be very dangerous. Safety must be the foremost priority of anyone attempting to fell a tree. Failure to follow the step-by-step instructions or use the proper Personal Protective Equipment (PPE) could result in serious injury or death. It is wise to never cut firewood or fell trees alone.

Disclaimer

Although tree felling can be an enjoyable task, it also poses serious risks and safety concerns. This technical instructional manual is in no way intended to replace any type of formal tree felling or logging training. Therefore, the author of this manual cannot and will not be held responsible for any type of injury or death that may occur during any or all of the processes explained in this manual.

Figure 6.1
Front Matter from Former Student Eden Dorsett
(continued)

Definitions

- ***Felling***—The process of cutting down a tree.
- ***Limbing***—The process of cutting off all limbs and branches of a fallen or standing tree.
- ***Bucking***—The process of cutting a felled tree trunk into useable sections or logs.
- ***Lean***—The natural direction in which the tree tends to lean towards. Lean also provides insight as to which direction the tree will fall.
- ***Landing Area***—This is the area in which the tree will land when it falls.
- ***Escape Path***—This is a clear path of escape on the opposite side of the tree leading away from the expected fall direction of the tree.
- ***Notch***—The notch is a V-shaped cut on the side of the tree that is intended to be in the direction of fall. This V-shaped cut is approximately a 70-degree angle.
- ***Hinge***—The hinge is the remaining wood attached between the stump and the tree trunk after the notch and back cut are made. This small section of wood that is still attached becomes the pivot point, when the tree is falling, between the tree trunk and the stump.
- ***Back cut***—The back cut is the final cut made immediately before the tree begins to fall. The back cut is a straight horizontal cut on the opposite side of the tree from the notch.

Introduction

What is your set of instructions about? Start your instructions with an introduction. People like to know up front if the set of instructions is written for them; they like to know what you're getting ready to tell them; they like to think they can succeed in accomplishing the task they're getting ready to do. Here are questions you need to answer in the introduction:

- What do the instructions cover? ("This manual walks you through the basics of how to set up a tropical fish aquarium.")
- What is the scope of the instructions? ("This manual covers only setting up a tropical fish aquarium; it does not cover . . .")
- Who is the audience? ("This manual is written for folks who have never . . .") See Figure 6.2.
- What assumptions are you making about the audience? ("This manual assumes you already know how and where to purchase the appropriate fish and fish tank.")
- What have you done to motivate the person to read the instructions? Have you shown that accomplishing the task is within the person's reach? Have you mentioned how and why the instructions will help them?

> These instructions are designed for the novice and experienced user. If you are a new user, you should start reading at the beginning of Section 1 and consult the Glossary when you encounter unfamiliar words. Novice and experienced users may search through the Table of Contents and use the headings throughout the manual to find things quickly.

Figure 6.2
Partial Sample of a Student Introduction to a Set of Instructions

Materials/Equipment/Tools Needed

What materials/equipment/tools does the reader need to accomplish the task? People get frustrated when they have to stop mid-job because they are missing a tool or part. So, give the reader a complete and precise list of materials and equipment needed! We have already discussed accuracy in this book. It is an important concept in instructions. Have you described and quantified the materials/equipment/tools? Have you included visuals of the materials/equipment/tools if the user will need that assistance?

Safety Information

What safety precautions does the reader need to know up front? It is critical that safety precautions be included if needed. Clearly describe what will happen to the user if the precautions are not followed:

- Frustration?
- Minor injury?
- Major injury?
- Death?

INCLUDE VISUAL CUES FOR READER

An effective way to communicate important messages is to use visuals to cue your reader. On this page, the skull and crossbones indicates *grave danger,* including *death*; the exclamation points acts as a *warning* or *caution* signal, indicating injury is possible; the electricity sign indicates a specific kind of *electrical danger*; the stop sign means just that, *stop*; and the notation sign indicates you are providing an *explanation or clarification* for the user's information.

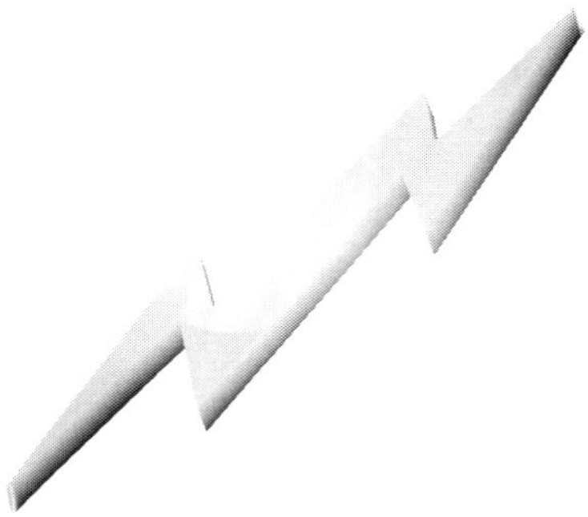

MIDDLE MATTER

The middle matter is the meat of your instructions. It is where you describe step by step what the user is to do.

Step-by-Step Instructions

What are the step-by-step instructions? Here's where your communication skills are tested. The level of expertise of the user will determine how you approach the middle matter. In your planning stage, you analyzed your user, so where on the continuum does your targeted user fall?

Novice ——————X——————————Expert

Here are important tips to help make sure your user reads and understands your instructions.

- Chunk the instructions under categories of steps. *Users can handle small chunks more easily than large chunks. In other words, don't overwhelm the user from the get-go!* See Figure 6.3.
- Number each step, starting with number 1 under each major chunk.
- Start out with what the user is getting ready to do and explain *why*. *Users like to know why they're doing something. Including the why immediately following the heading of a chunk is very important.* (See Figure 6.3.)
- Start each step with an action verb.
- Give only one action per step.
- Separate steps from explanation. *Use formatting effectively to distinguish the two. For example, you can italicize the explanation like we're doing here. Remember, whatever formatting tool you decide to use, you'll need to consistently use it throughout the instructions, allowing the reader to discriminate what she has to do from what she needs to know.*
- Use consistent standards throughout the instructions (headings, steps, white space, formatting, etc.).
- Use parallel structure throughout. *Notice how we use an action verb to begin each of the items in this bulleted list; that's parallelism.*

Section 1. Connecting to the Wireless Network

In this section, you will connect to the wireless network. After you complete this step, your laptop will automatically connect to wireless whenever you are on campus.

1.1 **Power on your computer and log into Windows.**
1.2 **Wait for the message in Figure 1.1 to appear** in the lower right-hand corner of your screen. If you do not see a message, your wireless card is not turned on.

Figure 6.3 Partial Student Sample of Effective Chunking

- Where needed, provide a visual for each step. *As stated earlier, visuals are a necessary part of instructions. Some users, visual learners, may rely almost solely on the visuals to perform the task, so your visuals must be accurate, clear, and large enough so the user can distinguish the finer points.*
- Place the written steps before the visual or to the left of the visual, depending on the format you're using.
- Refer to the visual in the step (e.g., "See Figure 3").
- Number and title all visuals (e.g., "Figure 3: Photo of a Goldfish").
- Use formatting techniques effectively throughout to make the information clear and available to the user (boxes, bold, italics, color, white space). *Keep these formatting techniques consistent throughout the set of instructions. For example, if you put one "Caution" in a box to call attention to it, do the same for all "Cautions" in the instructions.*
- Write clear, effective, concise, simple sentences.
- Use plain and precise language; avoid jargon.
- Write directly to readers (i.e., use *you*), which draws them in and lets them know you're writing the instructions for and to them.
- Make sure the instructions are accurate and complete. *Your usability testing discussed later in this chapter will help you with this one.*
- Make sure the instructions are error-free.

WHAT IS PARALLELISM?

Parallelism is using the same grammatical structure for items that are equal in value and degree. Parallelism is particularly important in a set of instructions because it helps guide the reader to know what to expect. In other words, by using action verbs to begin steps, you are providing the reader with a guidepost. See Chapter 8 for a more detailed discussion of parallelism.

BACK MATTER

The back matter provides closure for the reader. Typically, back matter includes a conclusion and troubleshooting and resources sections.

Conclusion

Did you bring closure for the user? Users need to feel they have completed the task successfully. Sometimes just a simple "Congratulations! You've just designed your first Web page" is enough. See Figure 6.4.

Troubleshooting

Did you tell users what to do if something doesn't work right? In an ideal world, nothing would ever go wrong, but things do go wrong. So let users know workarounds for possible problems or, if there is no workaround, at least a possible resource where they can find help. How did you present this information? Make sure you use a format appropriate for the user and the topic. Sometimes a table format is the best option.

Resources

Did you give readers a number to call or another print source to refer to if they want more information? Sometimes, readers like to research the topic a bit more before diving into the task, especially if it's a complex process or mechanism.

Figure 6.4 Student sample of an Effective Closing

> **Closing**
>
> You should now be able to successfully play a game of tenpin bowling. Always be sure to follow the safety instructions to avoid injury. It is important to obey common sense safety at the bowling alley and to avoid playing with the bowling ball outside of regular play, as an injury or property damage could result.

How to Make Callouts Using Microsoft Word 2007

A good way to point out an important part of a visual or to make a comment on a visual is to use a callout using Microsoft Word 2007. Here are simple instructions to make callouts.

1. Click on the "Insert" tab in your Menu bar.

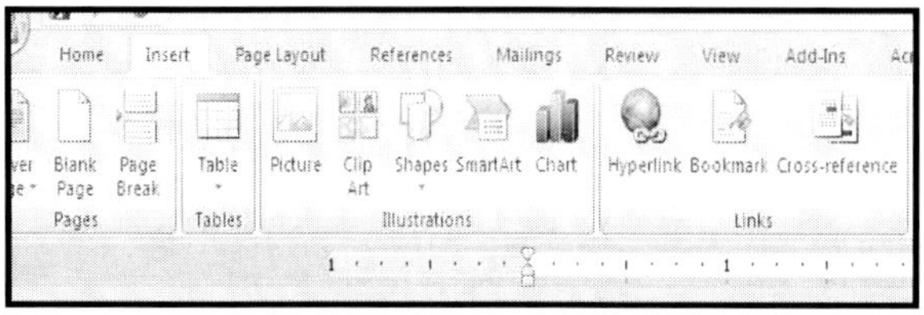

2. Click on "Shapes." *You'll see a dropdown box showing many different shapes.*

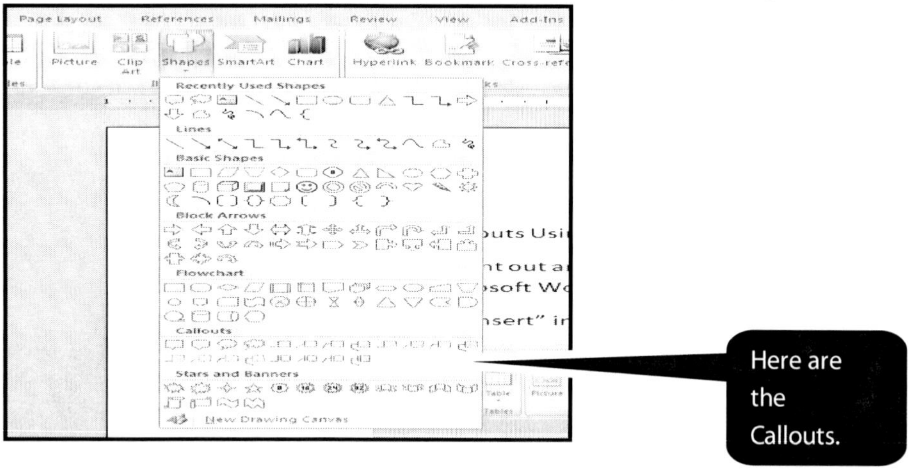

3. Select the type of callout you want to use.
4. Left click the mouse where you want the callout to appear.

A "sample cloud" callout shape

To type inside the callout, simply left click inside the shape and start typing. When your callout appears, the "Format" tab under Text Box Styles opens up automatically (see below). You can then format the callout to fit your needs and purpose.

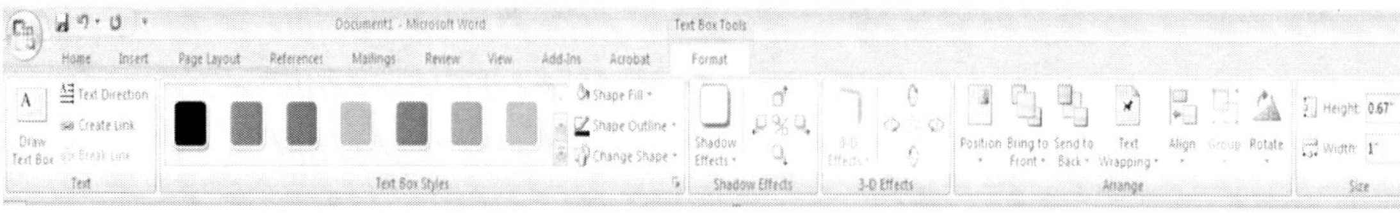

Usability Testing

> **Note:**
>
> What is obvious to you might not be obvious to the user.

Wouldn't it be a shame if you wrote a beautiful set of instructions and no one could follow them? Of course, it would. That's why we do usability testing on potential users to find out if users can follow our instructions. Begin with a detailed plan.

CREATE A PLAN

To create your usability testing plan, answer these questions:

- How do you plan to test your instructions? Will you need a questionnaire for users to complete as they perform the test? Or following the test? Will you observe users following your instructions and record your observations? Will you video record or audio record the user following your instructions?
- Who do you plan to test? Novices? Experts? Both? Strangers? People you know? Both? Remember to include users representative of your target audience.
- How many people do you plan to test?
- When do you plan to test?
- Where do you plan to test?
- What do you consider a successful test? For example, how much time should the user take to perform the test?

PERFORM USABILITY TESTING

When usability testing, it's paramount that you perform the testing consistently with each participant. In other words, you want to use the same procedure, use the same physical setting, ask the same questions, allot the same amount of time, for example. Doing so ensures the data you gather is authentic.

REVISE INSTRUCTIONS BASED ON RESULTS OF USABILITY TESTING

Once your usability testing is complete, you'll want to revise your instructions based on any problem areas the users found in your set of instructions. Perhaps a user didn't understand how to complete one of the steps. Hopefully, during the test, you were able to ascertain *why* the user had problems with the step. This information is invaluable as you revise your set of instructions to meet the users' needs.

Writing Process Strategies

If your project is a **class project** and you're having problems thinking of a topic to write about, make an ***authority list.*** See Table 6.1. An authority list is simple, really. It's a list of things you do well and things you know a lot about. So, to make your own authority list, do the following:

1. Brainstorm a list of things you do well. (Remember, when you brainstorm, don't censor yourself.)
2. Brainstorm a list of things you know a lot about.

Table 6.1 Sample Authority List

Things I Do Well	Things I Know a Lot About
1. Manage projects	1. How light emitting diodes work
2. Use computer-aided design software	2. Current construction practice and methodology
3. Calculate how far away lightning strikes by how long it takes for thunder to arrive	3. How UPC bar codes work
4. Rewire a house	4. Small-scale electronic systems
5. Change a tire	5. American history during the Civil War

An Authority List is simply a heuristic to start you thinking about possible topics. In the **workplace**, however, you rarely have the luxury of choosing the topic on which you'll write a set of instructions. More likely, a supervisor or colleague will ask you to explain *how to* do something to be shared with your colleagues at a staff meeting, for example. Or, maybe your supervisor asks you to document in writing how to complete the travel form and to share that with company employees.

Sometimes you may have all the information you need in your head; sometimes you'll need to do research.

If you're ready to begin, do this:

- Start by clearly naming your topic. *Review your authority list for possible topics, if a class project.*
- Analyze your audience! *Describe a typical user.*
- Review manuals you have at home or at work to get an idea of what manuals look like (e.g., automobile, copier, software package, Smartphone, lawnmower).
- Brainstorm, freewrite, or cluster on the topic of the instructions.
- Do your research. *Gather the information you need to write the manual.*
- Write a first draft of the manual.
- Review your draft and make changes.
- Get feedback from colleagues. *If in a highly regulated industry, your instructions will be subjected to rigorous review.*
- Revise using the feedback from your colleagues.
- Test your instructions for usability.

- Edit based on usability test results.
- Write the final draft of your manual.
- Proofread.

Here are two resources on usability testing you may find helpful:

1. Information & Design's Usability Testing Process (http://www.infodesign.com.au/usabilityresources/evaluation/usabilitytesting.asp)
2. Jakob Nielsen's Alertbox on Usability (http://www.useit.com/alertbox/user-testing-showbiz.html and http://www.useit.com/alerbox/20030825.html)

Sample Student Manuals

See Figure 6.5 for an example of a good student manual. See Figure 6.6 for a partial student manual (middle matter) annotated with instructor comments.

Figure 6.5
Sample Student Manual

NOTE: The authors are aware that newer versions of Norton Ghost are available and that floppy disks are no longer used, but wanted to share this student manual because it is well written and formatted.

Creating and Restoring a Hard Drive Image Using Norton Ghost 2003

Author: Patrick Rasche
Date: 9/17/2008

Figure 6.5 Sample Student Manual *(continued)*

Table of Contents

Figure 6.5 Sample Student Manual *(continued)*

Creating and Restoring a Hard Drive Image

This manual will walk you through the process of backing up your computer using Norton Ghost 2003. Norton Ghost 2003 is a software package that allows you to back up the information on your computer hard drive to a set of DVD's. This hard drive backup is also called a "Drive Image." You will be able to use this image to restore the contents of your computer when needed. Some computers come from the factory with a restore disk; others do not. Even if your computer was packaged with a restore disk when you bought it, you may want to make a new drive image after you install your software. This manual shows you how to make a backup of your computer with everything you have on it today. It also shows you how to restore that information back to your computer. You may need to restore your drive image if you have experienced a hard drive failure, corrupted software, or a computer virus. Norton Ghost 2003 will boot your computer in PC-DOS to make sure that no information is changing on your hard drive as you make the drive image.

Materials Needed

Personal Computer Running Windows 2000 or Windows XP
Norton Ghost 2003 Software
DVD Burner
Blank DVD's (number dependent on how much information is stored on your computer)
3½″ Floppy Drive
Blank 3½″ Floppy Disk

Section 1: Creating a Ghost Boot Disk

In this section, you will create the Ghost Boot Disk. This disk will be used to boot your computer in PC-DOS and create the drive image. This boot disk can also be used to restore the drive image if your hard drive has to be replaced.

> **Note:** If Norton Ghost 2003 is not installed on your computer, please refer to the software documentation for installation instructions.

1.1) Run Norton Ghost 2003 by clicking "Start" > "Programs" > "Norton Ghost 2003" > "Norton Ghost."
1.2) Click on "Ghost Utilities."
1.3) Click on "Norton Ghost Boot Wizard."

Figure 6.5
Sample Student Manual *(continued)*

1.4) Choose "Standard Ghost Boot Disk" and click "Next" as shown in Figure 1.

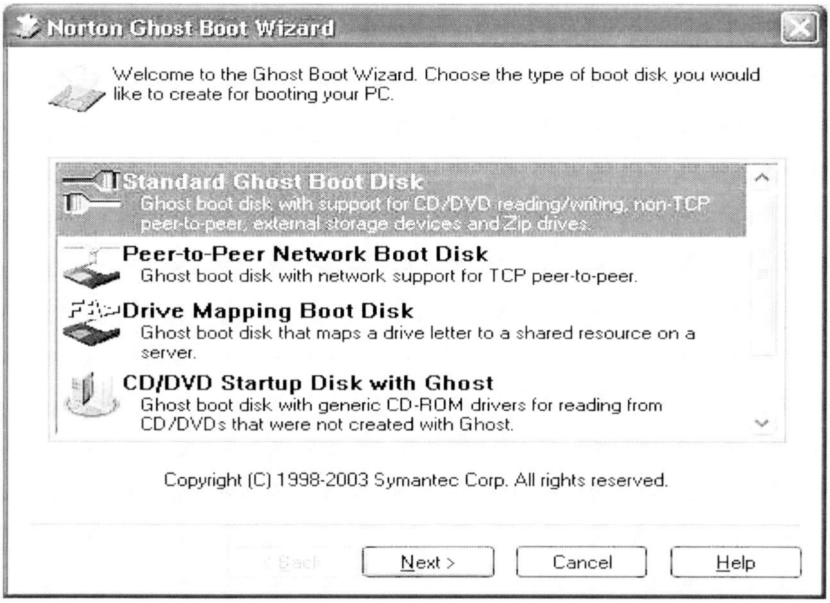

Figure 1: Norton Ghost Boot Wizard

1.5) Click "Next" on the "Additional Services" page as show in Figure 2.

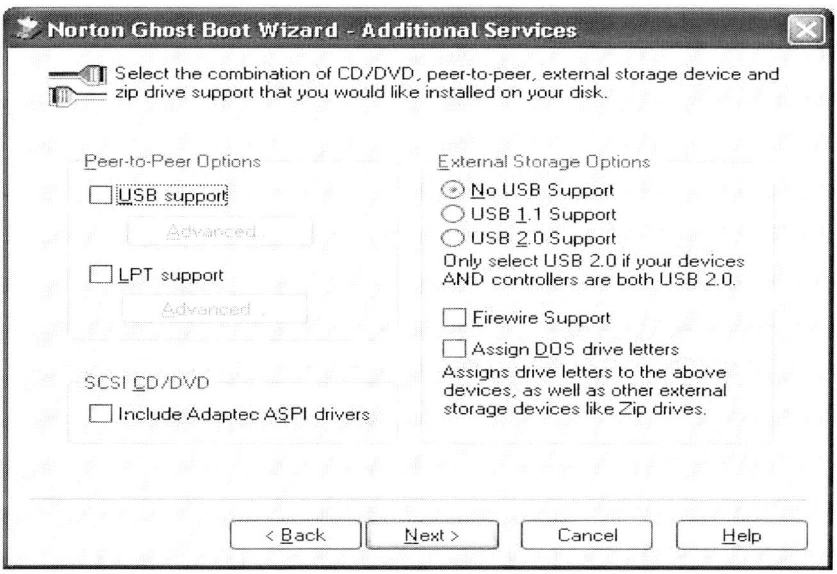

Figure 2: Additional Services

Figure 6.5
Sample Student Manual *(continued)*

1.6) Choose "Use PC-DOS" and click "Next" as shown in Figure 3.

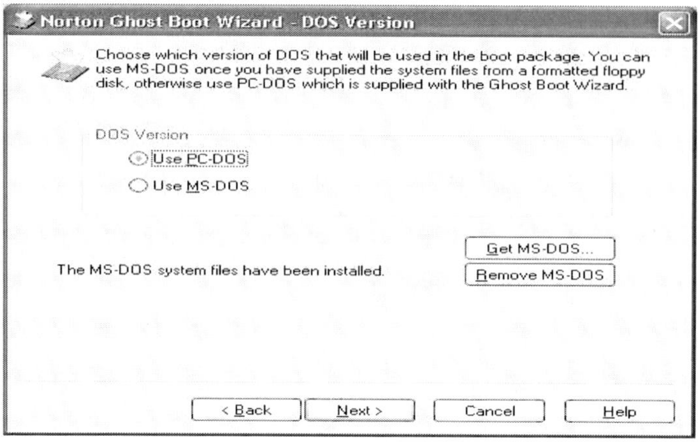

Figure 3: DOS Version

1.7) Insert a blank 3½″ floppy disk that you have labeled "Standard Ghost Boot Disk" in the floppy drive.
1.8) Select the drive letter you wish to save the file to, and click "Next." After pressing "Next," Norton Ghost will create your 3½″ floppy boot disk.

<u>Section 2: Creating a Drive Image</u>

In this section, we will create an image of all of the information on your hard drive. This image can be used to restore your data if you have a hard drive failure. This image will be stored on DVDs.

2.1) Shut down Windows using the Start menu. Turning the computer off by removing the power will cause errors in the image creation process.
2.2) Insert the 3½″ floppy disk labeled "Standard Ghost Boot Disk."
2.3) Start the computer using the power switch. The computer will boot using PC-DOS.
2.4) Choose "Local" > "Disk" > "To Image" as shown in Figure 4.

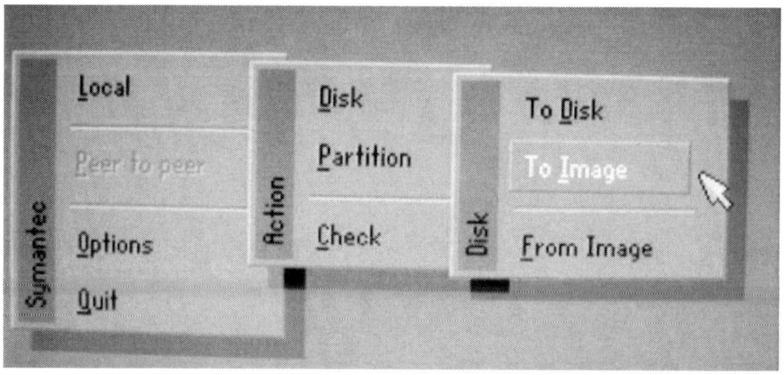

Figure 4: Disk "To Image" Selection

Figure 6.5
Sample Student Manual *(continued)*

2.5) Select the local source drive by clicking on the drive number.
2.6) Insert a blank DVD that you have labeled "Disk 1," in the DVD drive.
2.7) Use the drop down menu labeled "Look in," to select the CD/DVD drive. The file name will automatically change to "CDR00001.GHO."
2.8) Click "Save."
2.9) Click "Fast" when asked to choose a file compression method.

> **Note:** You will need to make sure the 3½″ floppy disk labeled "Standard Ghost Boot Disk" is still in the disk drive before proceeding.

2.10) Click "Yes" when asked if you would like to make the DVD bootable. The software will begin creating your disk image. You will be prompted if additional DVDs are required.
2.11) Click "Quit" when the drive image process is complete.
2.12) Click "Yes" when prompted to confirm that you would like to quit.

> **Note:** You will be returned to the PC-DOS prompt after quitting Norton Ghost 2003. You will need to use the power button to restart the computer.

Congratulations, you have successfully created a backup disk for your computer. If you make major changes to your computer, you can create a new disk image at any time. You can now feel safe that your data will not be lost.

Section 3: Restoring a Drive Image

In this section, you will restore the hard drive image to your computer. You will be using the set of DVDs that you created in Section 2.

> **Note:** You will not need to complete this step unless you have had a hard drive failure or you wish to restore your computer to the time when you made the backup.

3.1) Insert the first DVD of the drive image while the computer is running. This disk should be labeled "Disk 1."
3.2) Shut down Windows using the Start menu.
3.3) Start the computer using the power switch.
3.4) Choose "Local" > "Disk" > "From Image" as shown in Figure 5.

Figure 6.5 Sample Student Manual *(continued)*

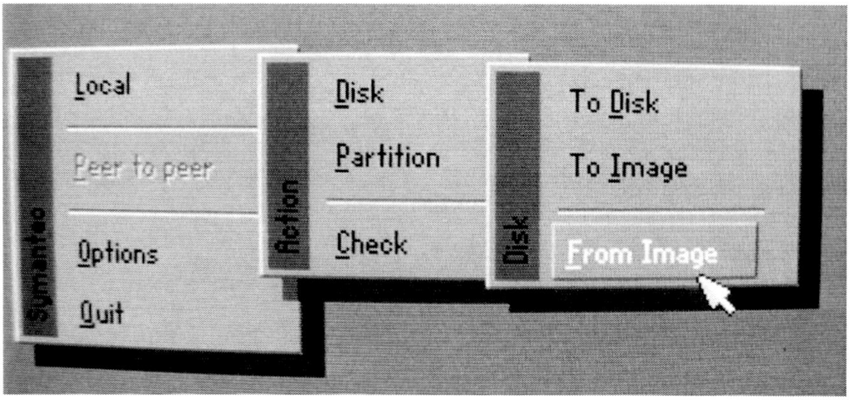

Figure 5: Disk "From Image" Selection

3.5) Use the drop down menu labeled "Look in," to select the DVD drive.
3.6) Click "Open."
3.7) Select the local destination drive as shown in Figure 6 and click "Ok."

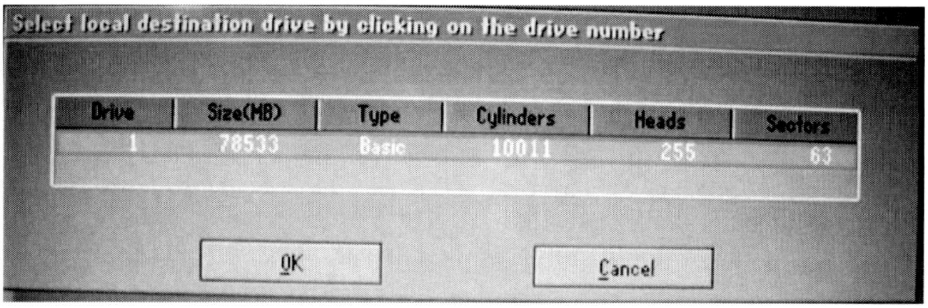

Figure 6: Local Destination Drive

3.8) Verify the destination drive details as shown in Figure 7 and click "Ok."

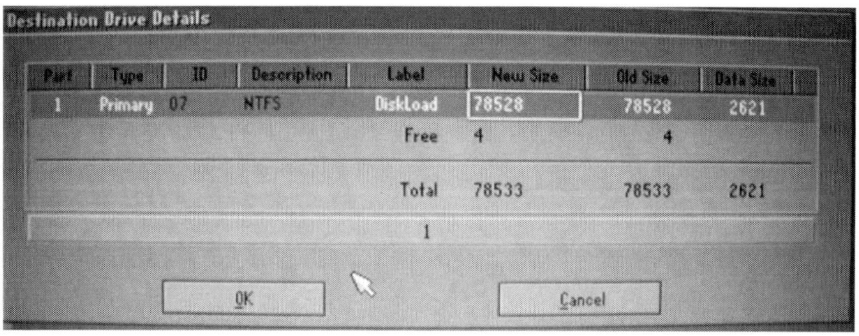

Figure 7: Destination Drive Details

Figure 6.5
Sample Student Manual *(continued)*

Warning: In the next step, you will be asked if you want to restore the image to your hard drive. If you choose "Yes," everything that is currently on your hard drive will be overwritten.

3.9) Click "Yes" to proceed with the "disk restore" as shown in Figure 8. The software will begin restoring data from the first disk.

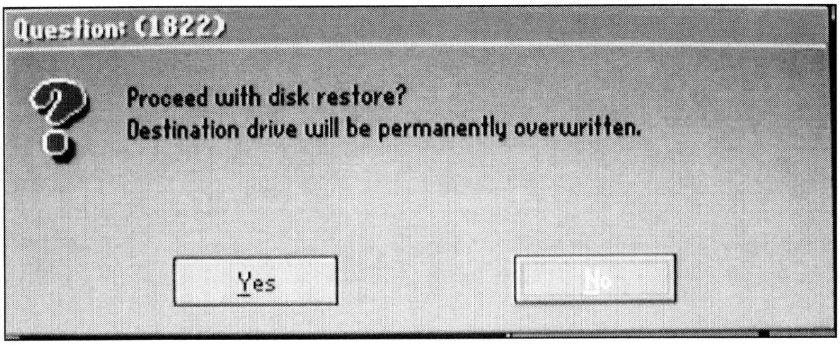

Figure 8: Drive Restore Warning

3.10) Insert the next drive image DVD if you have more than one in your backup set.
3.11) Click "Reset Computer" as shown in Figure 9.

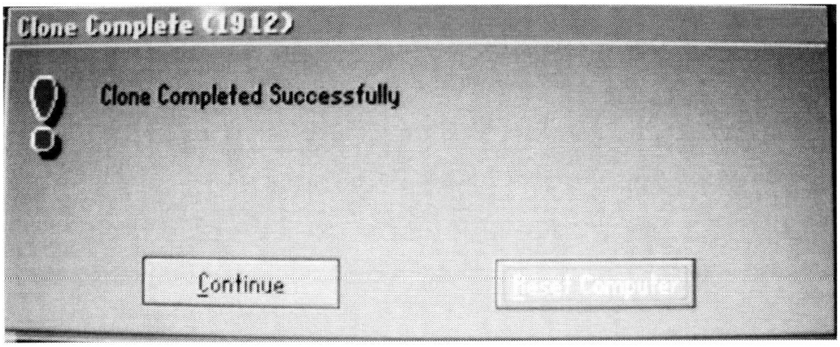

Figure 9: Reset Computer Message

Congratulations, you have just successfully restored your drive image to your computer.

Section 4: Troubleshooting

If the computer is shut down using the power button while it is running Windows, the following error may appear during the image creation process:

"Encountered an NTFS volume with a Logfile that has not been flushed. (536)"

Figure 6.5
Sample Student Manual *(continued)*

To correct this problem, you will need to restart the computer and complete the following procedure.

4.1) Click "Start" > "Run."
4.2) Type "command" in the run box.
4.3) Type "chkdsk/f" when the MS-DOS command box appears.
4.4) Press "Y" to indicate you would like the disk to be checked the next time the system restarts.
4.5) Restart the computer using the "Start" menu. After chkdsk is finished running, Windows will continue to load.
4.6) Go to step number 2.1 of the drive image creation instructions after Windows has finished loading.

Section 5: Conclusion

Congratulations, you have learned how to create a ghost boot disk and a drive image as well as restore a drive image.

Copyright notice: Norton Ghost and Norton Ghost 2003 are registered trademarks of Symantec Corporation. PC-DOS is a registered trademark of IBM Corporation. Windows, Windows 2000, Windows XP, and MS-DOS are registered trademarks of Microsoft Corporation.

Partial Student Manual (Middle Matter)

6. Line up the spring into its proper place and bolt the spring clip back onto the bottom of the spring.
7. Reattach the sway bar links on both sides.

Section 4.3: Front Shock Absorber Removal and Installation

1. In the previous section, you removed the bottom shock mounting bolts. Now you need to remove the nut holding the top of the shock on. Open the hood of the vehicle and place vise grips of the two flat spots on the shock absorber stem to keep the shock from turning. Using a boxed-end wrench, remove the nut and pull the shock out from the wheel well.

> **Comment [B1]:** This step actually has many different commands in it. Be careful not to load too many steps in what looks to be one, as that is a good way to lose your user.

2. First extend the new shock, then place the new washer and rubber bushing on the stem and push it into the upper mount. Then you need to tighten the stem nut securely.
3. Put the lower end of the shock into position and install the lower mounting nuts and bolts, tightening them securely.
4. Slide the tires back onto the vehicle and snug up the lug nuts. Lower vehicle off the jack stands. Finish tightening the lug nuts.

Section 4.4: Steering Damper Replacement

1. First remove the nut securing the steering damper to the center link.
2. Next, unbolt the damper from the frame of the vehicle.
3. ~~Now you will need to~~ strike the bolt that is holding the damper to the center link to free it up from the center link. Then remove it from the vehicle.

> **Comment [B2]:** You do not need introductory "now you will . . ." verbiage because you have numbered the steps. Just proceed with the next verb.

4. Put the bolt and wedge on the new damper and attach it to the center link. Do not fully tighten the nut.

> **Comment [B3]:** Again, be aware of two steps in one!

5. Extend damper to mounting point on the frame and start that bolt and nut.
6. tighten both bolts up to finish installation.

Section 4.5: Rear Leaf Spring Removal and Installation

1. Do the next steps for both sides of vehicle.
2. Break the lug nuts free on both back tires. Chock the front tires. Raise and place securely on jack stands. Place the jack stands just in front of the front leaf spring mounting bolts. Place floor jack under rear axle to support it after springs are removed.

> **Comment [B4]:** Another example of multiple steps in one. Compare this step to the ones above, where you were careful to keep each individual step distinct.

3. ~~Now you will need to~~ remove the rear tires. Unbolt the shock from the axle bracket.

> **Comment [B5]:** Start with verb.

Figure 6.6
Partial Student Manual (Middle Matter) annotated with instructor comments

Figure 6.6
Partial Student Manual (Middle Matter)
Note annotation comments from instructor.
(continued)

4. Unscrew the u-bolt nuts and remove the spring tie plate.
5. Remove the rear spring to shackle bolt. Then remove the front bolt holding the spring to the frame. Remove the spring from the vehicle.
6. Using the c-clamps, clamp the leaf springs together to keep them from coming apart. Using the propane torch, heat up the band clamp holding the smaller leaf spring on, then remove the center eye bolt and slowly release the center c-clamp. Remove the small leaf spring and place the helper leaf that came with your lift kit between the next size bigger leaf and the smallest leaf.

Comment [B6]: Three removals in one step!

7. Line up the center hole and place new center eye bolt in the spring and start the nut. Compress the springs together with the c-clamp, not the bolt. Reform the band clamp around the newly compressed springs.

Comment [B7]: Break out the steps.

8. Allow to cool before going any further.
9. When cool, remove the c-clamps and slide springs back onto the vehicle. Start the front spring bolt. Then start the rear shackle bolt.
10. Place tie plate back on spring and using the new U-bolt and nuts, begin securing springs to axle. Tighten u-bolt nuts to proper torque. Do not tighten the front and rear leaf spring bolts until the car is sitting on the ground and off the jack stands.
11. Proceed to next section to replace the shock absorbers before finishing the leaf springs.
12. Put tires back on vehicle. Lower off the jack stands. Finish tightening the lug nuts.
13. Now tighten the front and rear mounting bolts for the leaf springs to their proper torque.

Section 4.6: Rear Shock Absorber Removal and Installation

1. From the previous section, you unbolted the bottom shock bolt from the axle bracket.
2. Remove the two upper mounting bolts and remove the shock from the vehicle.
3. Place lower end of shock on the axle bracket and start the nut.
4. Extend the shock and start the top mounting bolts.
5. Tighten both lower and upper bolts to proper torque.
6. Return to previous section to complete leaf spring installation.

Comment [B8]: You have done a good job of separating the main steps, but you tend to put multiple steps in each instruction. Keep in mind a new user might be struggling to read and do at the same time, and you want to make your instructions as simple to follow as possible.
These instructions are certainly shorter and less involved than those in your previous section.

Chapter Summary

Writing an effective set of instructions can be challenging, even for experienced writers. Writing an ineffective set of instructions can be devastating to your company or organization. Analyzing who will be using your instructions is the first step. This chapter has given you many tips on how to plan, write, and test your instructions.

End of Chapter Activities

ACTIVITY 1 Writing Instructions

Write a set of instructions that is easy to read and follow. Include well-designed and well-placed visuals to help users understand how to complete the task. In addition, use appropriate language and design for the audience and the purpose (length: minimum five pages).

ACTIVITY 2 Usability Plan and Tool

Design a usability plan and tool for a set of instructions you've written on one of the following topics:

1. How to tie a tie
2. How to change an electrical light switch
3. How to clean the lawn mower
4. How to design a Web page

ACTIVITY 3 Results of Usability Testing

Report on the results of usability testing you performed on a set of instructions you wrote. Use memo format and include the following topics:

1. Overview of project
2. Test procedures (e.g., The number of people you tested your instructions on and their relevance to the material. Were they peers, fellow employees, or total strangers? A description of the process used to test your instructions.)
3. Test results (How successful or unsuccessful were your participants in following your instructions?)
4. Analysis of results (What do your results mean?)
5. Plans for revisions (How did/will these results affect your revision?)

ACTIVITY 4 Manual Analysis

Find two or three manuals to bring to class. Before class, jot down the things that work well in the manuals and the things that don't work well. Once in class, form groups of three to five. Each group member shares her/his list with group members. As a group, make a list of characteristics of effective manuals. Be prepared to share your list with the class.

ACTIVITY 5 Group Practice Writing Instructions

Form groups of three to five. Use the items provided by the instructor to construct a mechanism. Write instructions on how your group put the mechanism together. Use the advice given in the chapter. Deconstruct the mechanism and trade your items and instructions with another group in the class. Each group will use the instructions to construct a second mechanism. Each group will describe how well (and why) it was able to follow the classmates' instructions. Be prepared to share your experience with the class.

CHAPTER 7

Data and Information Sharing Strategies Using Graphics/Visuals

Overview

Beyond the content of our message, the way we visually display the information often determines whether the audience receives the message that we intend. This display includes everything from formatting elements (such as text spacing, headings and subheadings, font style or size, and white space) to inclusion of graphics (such as tables, figures, maps, illustrations, drawings, and photographs). We rely on our knowledge of the conventions of a particular type of document to tell us what we are looking at. For example, the formatting elements of an e-mail are different from those of a formal business letter. Because we are familiar with the formatting conventions, when we see an e-mail, we know it is an e-mail.

We visually display or design our messages to help the audience access and understand them. For example, we use headings and subheadings to help the reader know where to find certain information. We place information in a certain order to help the reader know what is more important in our message. We use formatting such as bold to emphasize a word. So, how we visually design our message helps determine the audience's experience with that message.

In the workplace, we are bombarded with information and numbers, or *data*, as we refer to them in this chapter. How do we as writers organize the data so that we can work with it logically—so that we can draw conclusions, make recommendations—and so that our readers can use it to draw conclusions, make recommendations, and make informed decisions? Displaying this data in visual form helps make what are sometimes random or abstract pieces of information more concrete for the reader. Dense, complicated data, whether text or numerical, are much easier to understand when organized and presented visually in tables and graphs.

So, organizing raw data—especially complex numerical raw data—in a systematic way goes a long way in helping you and your reader understand it. In fact, raw data are often nothing more than a bunch of numbers or words to the reader if not organized and discussed, and then displayed in an effective visual.

At this point, we have used the word *data* often in this chapter, but how exactly do we define it? *Data* are "individual facts, statistics, or items of information," a "body of facts or information" (retrieved on March 15, 2010, from http://dictionary.reference.com/browse/data). *Data* are also a "single piece of information, as a fact, statistic, or code; an item of data" (data. (n.d.). Dictionary.com Unabridged. Retrieved March 15, 2010, from Dictionary.com Web site: http://dictionary.reference.com/browse/data).

So, for our purposes, *data* refers to statistics, maps, pictures, diagrams, facts, any information we use to help us draw conclusions, make recommendations, and make decisions. Visually seeing how one item relates to another or how an item changes over time or how an item changes when new variables are added helps your reader comprehend the meaning of the data. In other words, by visually presenting the data in tables and charts, you make the information more accessible to the reader. The reader can then use the data to make decisions.

PURPOSES OF VISUALS INCLUDE:

- Increasing utility
- Emphasizing important data and information
- Increasing aesthetic appeal of the document
- Increasing the persuasive impact of your point
- Making the abstract more concrete
- Condensing large amounts of data

Tables and graphs, modes of visual communication, are prevalent in workplace technical communication. Visuals are a way to emphasize data. However, visuals can't just be dropped into a report; they must be referred to and discussed. Readers appreciate having the information presented in a visual explained to them. And the data included must support the conclusions and recommendations presented in the document, which means data must be accurate and complete. Readers also appreciate having to read only necessary data in a report, so the data included must be necessary.

As with all communication, ethics play a huge role in visual communication. As the writer, you have the responsibility of presenting data accurately and completely to your readers.

This chapter introduces you to the simple basics of using tables and figures (graphs, charts, photographs, drawings) in your technical documents to communicate data visually.

A FEW TERMS TO KNOW

Visual rhetoric: the way images and text work alone and together to create a specific reaction in the reader.

Communication design: how design and information-development work together to communicate with the audience.

Visual communication: communicating through visuals; includes animation, film, digital images, signs, drawings, figures, illustrations, maps.

Information design (or document design): displaying information effectively and visually; includes logically organized information, emphasis of important information, clear writing, navigational clues, page design (font style, size, color, white space, etc.).

(See references at the end of the chapter.)

Ways to Present Data

You've considered your reader and purpose and decided to use visuals to supplement your text. The question now is what kind of illustration will best communicate the data. Again, you must return to analyzing the context of your communication: the data itself, your purpose, and the reader.

Does the information lend itself to a table or graph? Would a photograph or a drawing be effective? Would an organization chart help communicate your point? Is your purpose to summarize data? If so, then a table will work nicely. Is your purpose to emphasize data? If so, then a graph, drawing, chart, or photograph will work well. Sometimes you will want to start with a table of raw data and then use a graph to illustrate a particular point shown by the raw data in the table. Who is your reader? Does your reader relate better to both words and visuals? Is your reader international? If so, consider color and the type of visual carefully. Different cultures use different conventions to communicate visually!

Graphs are referred to as figures in the text.

The bottom line: Make all visuals simple and clear! Visuals should be easy to read and understand. They almost always supplement the text, but sometimes replace it (e.g., safety flyers on many airlines). They should include only essential data and should be carefully planned.

Tables

Tables are commonly found in technical reports. Tables use columns and rows to organize and visually display data. They provide a familiar way for you to share your information with the reader. Because the data are contained in a confined area, tables are made up of data cells. Sometimes the walls of these cells are visible; sometime not. You will decide which presentation is best for your readers, purpose, and data.

In addition, tables have column headers and row headers. Instead of data, these header cells contain descriptors that help readers know how to read and interpret the data in the data cells. Finally, tables should be numbered and titled. Place this label on top of the table.

Tables are used to present quantitative and qualitative data and allow you to do the following:

- Present large amounts of data in a small space
- Reduce text that otherwise would be dense with numbers
- Show exact numerical values
- Show relationships between/among ideas

See Figures 7.1 and 7.2 for sample tables, one that displays quantitative data only and one taken from former student Andrew Hoblitzell's report that displays qualitative and quantitative data.

Figure 7.1
Sample Table

Table 7.1 Errors for Younger and Older Groups by Level of Difficulty

	Younger			Older		
Difficulty	**M**	**SD**	**N**	**M**	**SD**	**N**
Low	.05	.08	12	.14	.15	18
Moderate	.05	.07	15	.17	.15	12
High	.11	.10	16	.26	.21	14

Figure 7.2
Sample Table

Table 7.2 Tasks and Estimated Costs

Task	Estimated Cost
Gathering data	Approximately 4 ($160) working hours
Performing analysis	Approximately 2 ($80) working hours
Evaluating the findings	Approximately 2 ($80) working hours
Implementing a trivial alarm system	Approximately 4 ($160) working hours and $5,000 for computer security enterprise alarm system
Implementing security solutions to the liking of ABC Corp.	Variable

Source: Andrew Hoblitzell, student

Graphs

Graphs come in many forms; the pie, bar, and line graphs are the most common. Graphs show:

- Relationships of two or more sets of data
- Absolute values or percentages
- Change over time

PIE (CIRCLE) GRAPHS

Data in a pie graph will equal 100 percent. Pie graphs are used to show relationships of parts of a whole and to show percentages and proportions. For example, Figure 7.3 shows the percentage of majors in the School of Engineering. It's best to limit the number of items in a pie graph to five or fewer and order the items from large to small, beginning at 12 o'clock. See Figures 7.3 and 7.4.

BAR GRAPHS

A bar graph is used to show relationships among wholes. A bar graph uses solid horizontal or vertical bars; each bar represents one item of information. Colors or patterns are used to help the reader see the individual relationships. See Figures 7.5 and 7.6. For example, Figure 7.5 displays the money spent for travel, for office expenses, and for salaries/benefits by quarter.

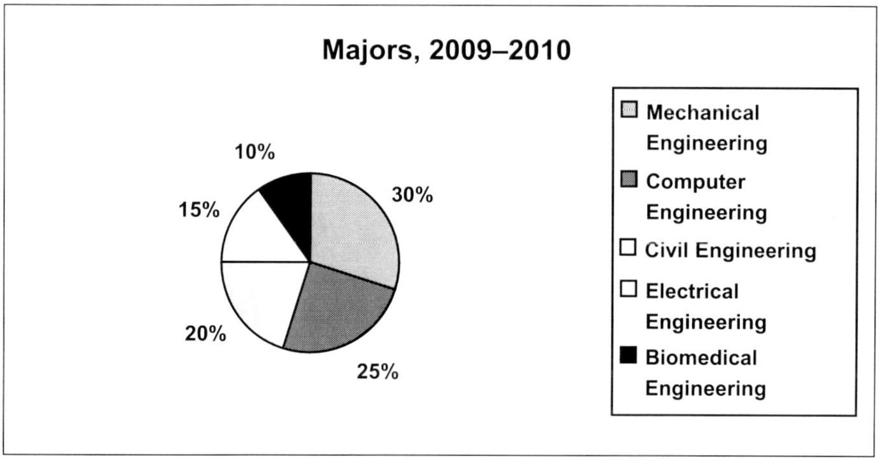

Figure 7.3
Sample Pie Graph

Figure 7.4
Sample Pie Chart

Source: *http://www.whitehouse.gov/stateoftheunion/2006/energy/index.html*

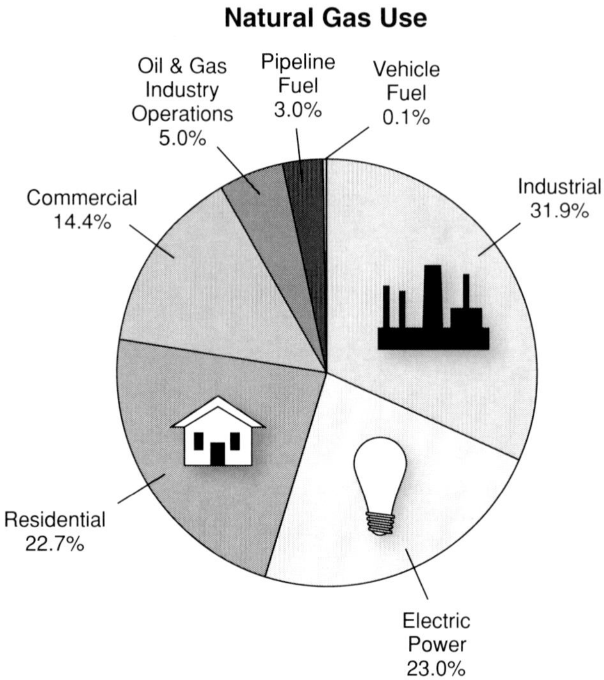

Figure 7.5
Sample Bar Graph

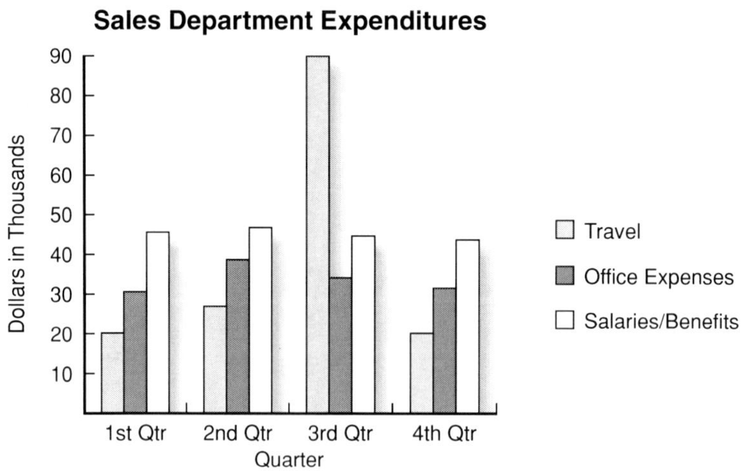

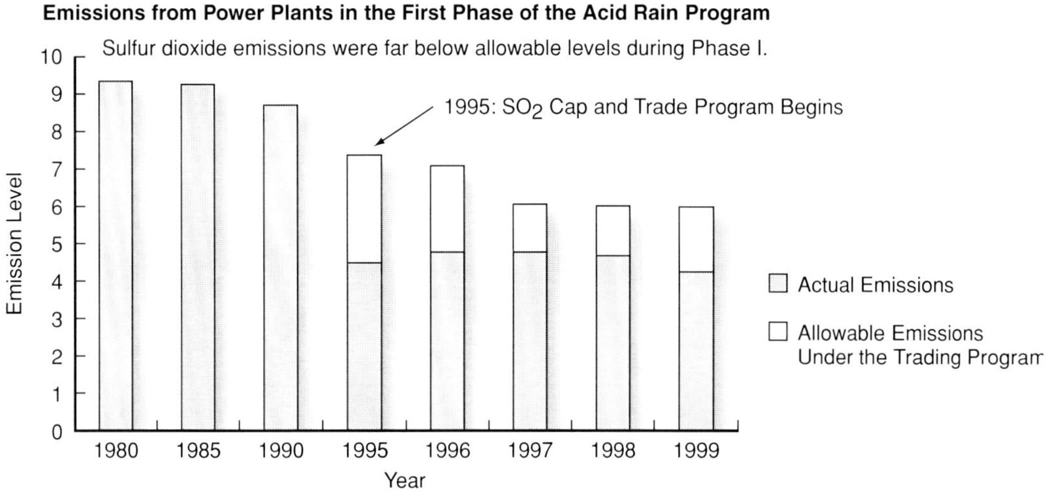

Figure 7.6
Sample Bar Graph

Source: *http://www.whitehouse.gov/news/releases/2002/02/clearskies.html*

LINE GRAPHS

A line graph shows the relationship of two quantitative variables, very often how a phenomenon changes over time. For example, Figure 7.7 displays how the population of the US has grown over a period of six years. Generally, the independent variable goes on the horizontal axis (x-axis) and the dependent variable goes on the vertical axis (y-axis). Having too many data lines on one graph will be confusing to the reader. The data should be easy to read.

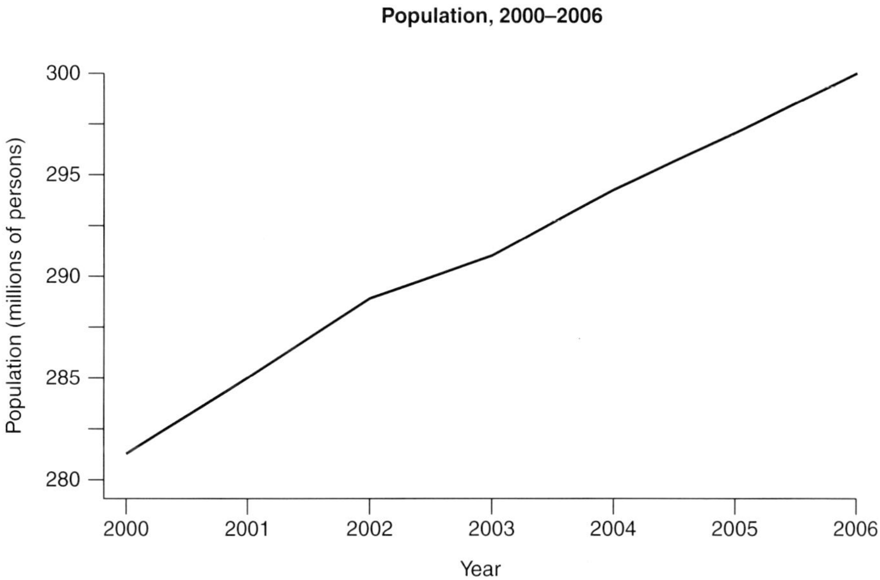

Figure 7.7
Sample Line Graph

Source: *http://www.gpoaccess.gov/eop/2007/B34.xls*

ETHICS OF GRAPHS

As with everything—and particularly so with numbers—you can skew your graphs to achieve your own ends. It is your responsibility as a writer of technical information to always accurately and honestly present the data! For example, a 3D pie graph can easily distort data by making those pie slices closest to the front appear larger than those farther away. See Figure 7.8.

Figure 7.8
Distorted Pie Graph

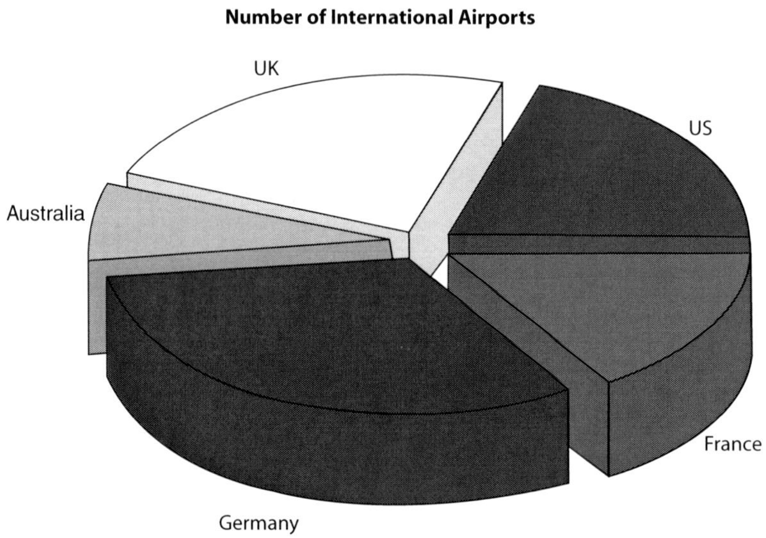

Charts

Common charts are flow charts and organization charts. These charts are most often displayed by using boxes (representing data) connected with lines. In general, flow charts show the sequence of steps in a process, whereas organization charts show hierarchy in a group. For example, you might choose to show the process of completing a job from start to finish in a flow chart or show the hierarchy of your company's management team in an organization chart. See Figures 7.9 and 7.10 for an example of each.

Photographs

A photograph is an easy and convenient way to show the actual appearance of something. It adds realism to your document and can show features that are difficult to draw or diagram (see Figures 7.11, 7.12, and 7.13). However, photographs also have limitations. For one, a photograph may include too much data, making it difficult to understand (see Figures 7.11 and 7.12). Also, a photograph illustrates only what the camera captures; you may need to see the item from an angle impossible for a camera to capture. In addition a photograph may be deliberately distorted.

Drawings and Diagrams

Like photographs, drawings and diagrams show the actual appearance of an object or life form; however, drawings and diagrams can show views of the object or life form that would be impossible to show in any other form. See Figures 7.14 and 7.15.

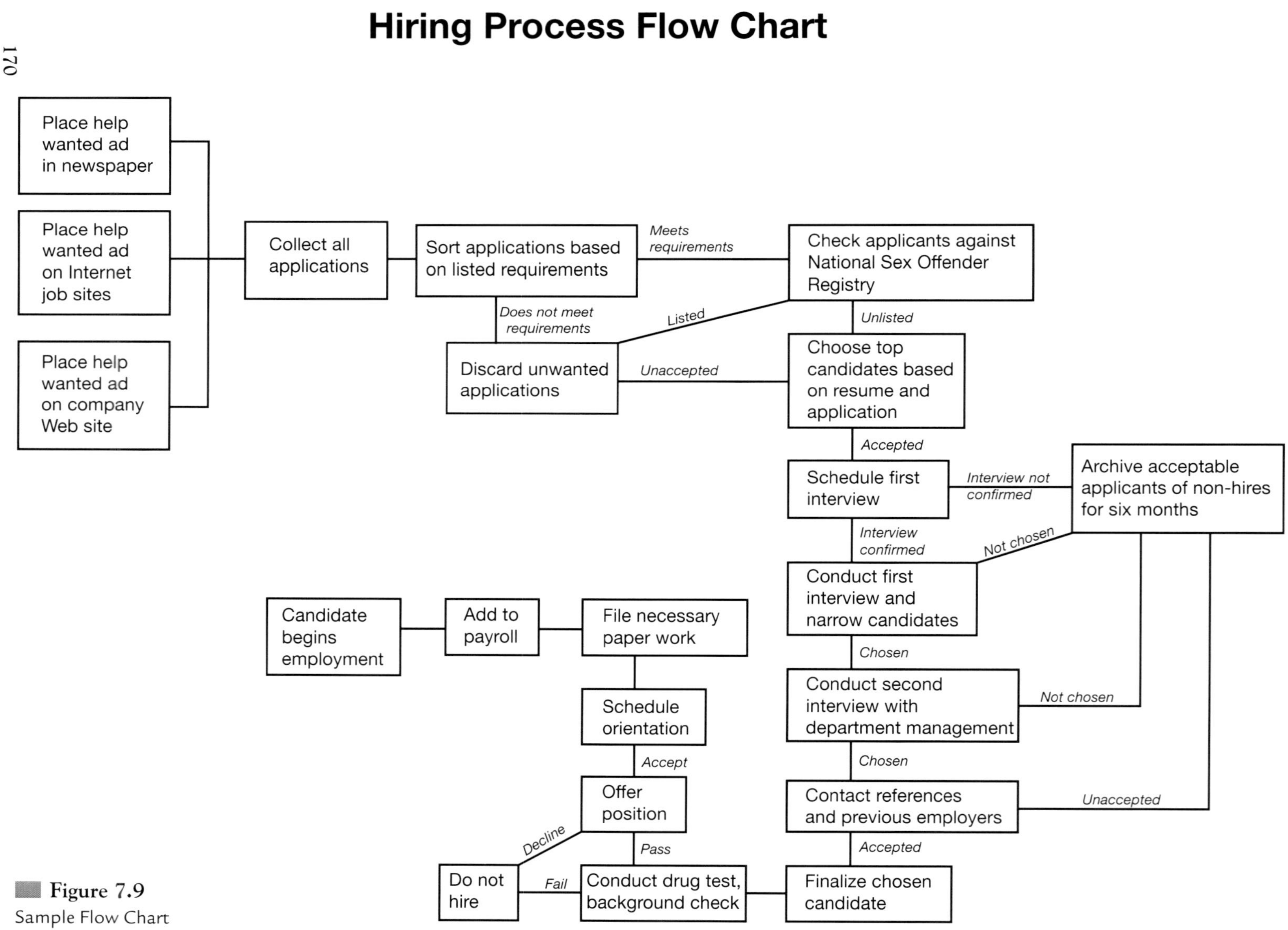

Figure 7.9
Sample Flow Chart

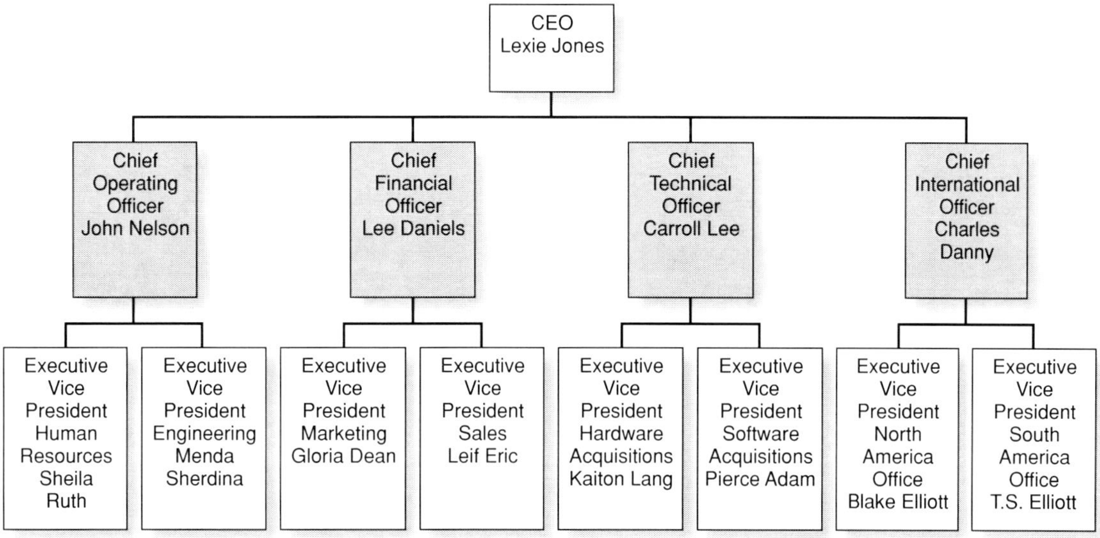

Figure 7.10
Sample Organization Chart

Figure 7.11
Construction Photograph

Figure 7.12
Nuclear Power Plant Photograph

Figure 7.13
Photograph of a Wind Mill

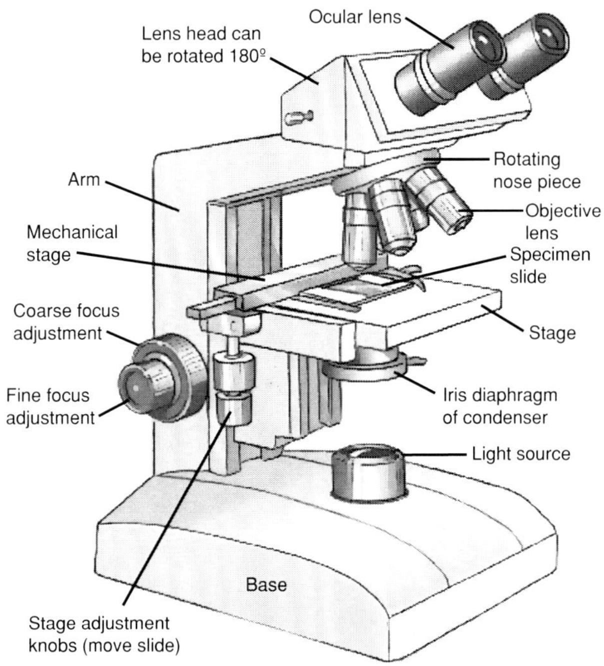

Figure 7.14
Sample Drawing

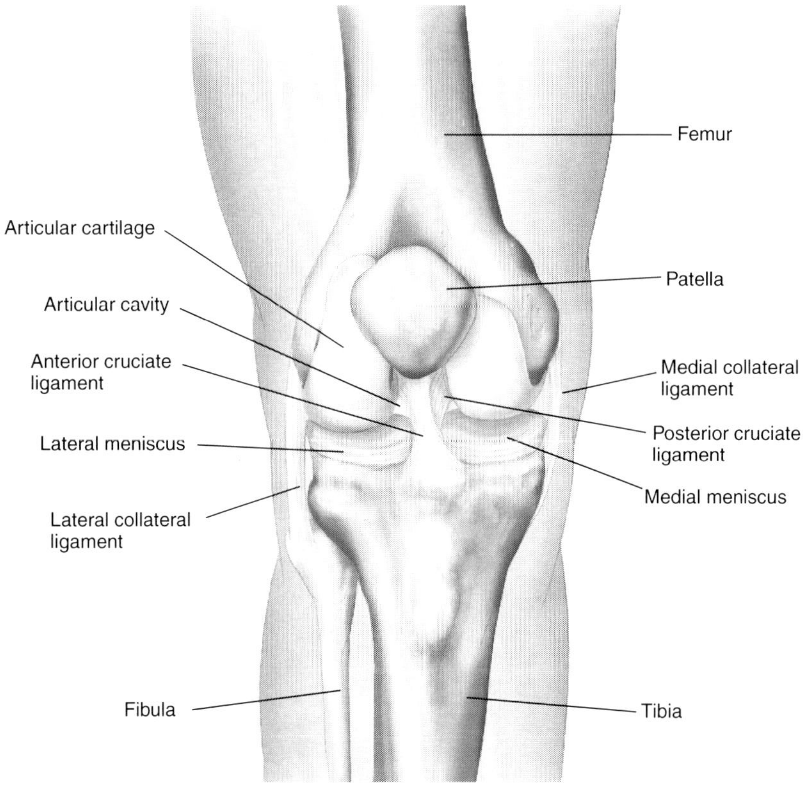

Figure 7.15
Sample Diagram

Screen Shots

Screen shots are commonly used in PowerPoint presentations and in instructions or manuals. Microsoft Windows makes it easy to make a screen shot. Simply press the **Print Scrn** key and make a screen shot of the entire screen you're currently viewing on the desktop. If you want a screen shot of the active window only, hold down the **Alt** key while pressing **Print Scrn.** (See Figures 7.16 and 7.17 for an example of each.) Although you will see nothing happen, your image has been stored on the clipboard. Once you are in the document (manual, report, e-mail, for example) where you want the screen shot to appear, click "Paste" and your screen shot will appear. You can also save the screen shot in Microsoft Paint (Windows' standard image editor) by selecting "Edit/Paste" if you think you might want to keep the visual to use in other applications. In Microsoft Paint, you can also edit the image if needed.

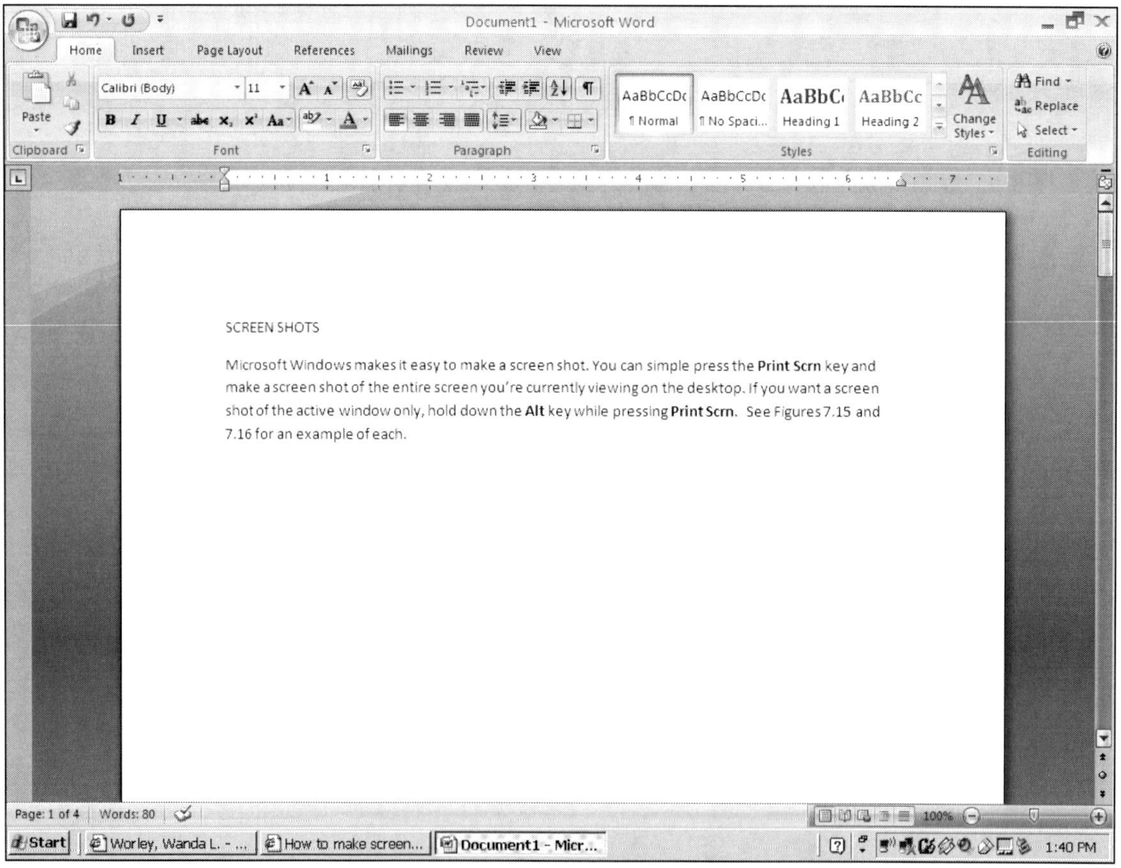

Figure 7.16
Screen Shot of Entire Screen

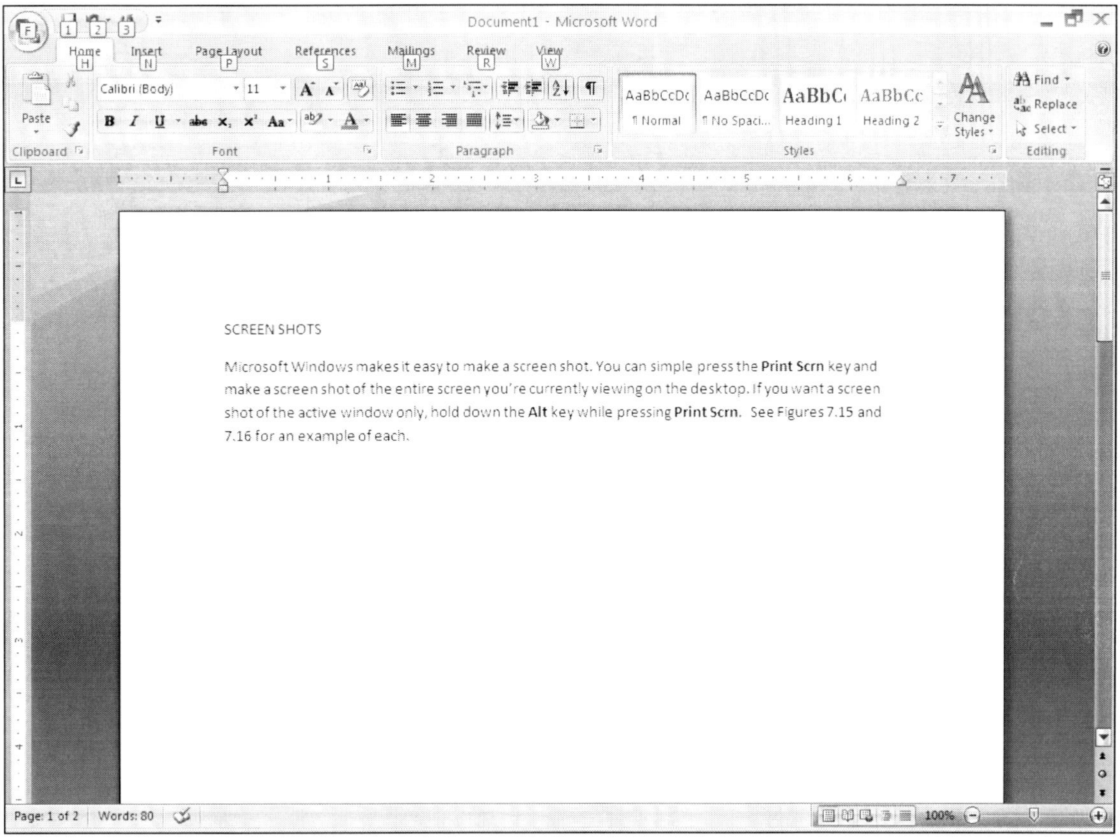

Figure 7.17
Screen Shot of Active Screen Only

For better results, you might want to use a screen capturing software. SnagIt by TechSmith is one of the most comprehensive and best known screen capture programs available. However, many programs are available, some free. A little research will help you know which one works best for your needs.

Integrating Visuals into a Document

Mentioned earlier in this chapter, visuals should not be dropped into a document without explanation; you must prepare your reader to read and understand them. In the United States, preparing your reader means you refer to the visual before sharing the visual. You will want to explain the data shared in the visual as well. Place your visual as close to the explanation as possible. However, if the visual interrupts the flow of the document, you can place it at the end of the document in an appendix.

Label (title and number) your visuals and cite the source beneath the label. Here's an example:

> **Figure 1:** Computer Technology Jobs
> *Source:* Smith, 2004, p. 9.

If you've labeled your visuals, you can easily refer to them by number in the text as is done in this chapter.

Chapter Summary

Tables, graphs, charts, photographs, and drawings are excellent tools for you to use in your technical documents to make raw data more understandable to your reader. Remember, though, consider the kind of data, your purpose, and your audience in determining how to display the data visually. Remember also to interpret and describe the data to the reader.

References

Carliner, S. (2000, fourth quarter). Physical, cognitive, and affective: A three-part framework for information design. *Technical Communication, 47*(4), 561–576.

Horn, R. E. (1999). Information design: Emergence of a new profession. In R. Jacobson (Ed.), *Information Design* (chapter 2). Cambridge, MA: MIT Press. Retrieved March 18, 2010, from http://www.stanford.edu/~rhorn/a/topic/vl%26id/artclInfoDesignChapter.html.

Kimball, M. A., & Hawkings, A. R. (2008). *Document design: A guide for technical communicators*. Boston: Bedford/St. Martin's.

End of Chapter Activities

ACTIVITY 1 Group Graphics Practice

Form groups of three to five. Pretend your group has written a report on the demographics of the student body in your school that work full- or part-time. Make up the data (or use the data provided by the instructor), and using Microsoft Word or Excel, design at least one visual that you'd use in a presentation to the Student Affairs committee. Be prepared to present your visual to the class and to explain to the class why you chose the kind of visual you chose.

ACTIVITY 2 Graphics Analysis

Examine several technical publications, preferably in your field of study (can use the Internet). Analyze the graphics used in these publications. Choose two or three graphics that are effective and two or three graphics that are not effective. Share what you learned about graphics in an e-mail to your instructor. Be prepared to discuss in class.

ACTIVITY 3 Ethics

Examine several technical publications, preferably in your field of study (can use the Internet). Analyze the ethics of several graphics. How would you revise them to be more ethical? Choose one and revise. Be prepared to present and discuss your revised graphic to the class.

ACTIVITY 4 Internationalization

Research how graphics are designed in other cultures. Use the databases you can access through the university library to complete your research. Using PowerPoint, prepare a brief presentation for the entire class on what you learned.

Find several graphics in professional journals in your field of study (can use the Internet). Analyze one of the graphics and note any items that need to be changed to fit an international audience.

ACTIVITY 5 Documenting Visuals

Draw a three-column table.

- Label the first column *Visual*
- Label the second column *Visual Location*
- Label the third column *Reference Location*

Locate and list all the visuals in Chapter 7 in a three-column table.

In the first column, list the number and title of each visual. In the second column, record the location (page and paragraph numbers) where each visual is located. In the third column, record the location (page and paragraph numbers) where the visual is referenced.

Pair with a classmate and discuss what you observed about placing, labeling, and referencing visuals.

ACTIVITY 6 Framing Visuals

Select an image or photograph or illustration to use in your document. Practice cropping it in various ways to see how the meaning or emphasis changes. Now practice placing the cropped versions of the visual in your document. Vary the size and placement of the visuals.

Pair with a classmate and discuss what you observed about the importance of cropping, placing, and sizing of visuals. How did the meaning of your message change?

CHAPTER 8

Writing Style

Overview

This chapter focuses on the characteristics of writing style. You have a writing style, but is it the style you want? Does it work for your purpose and audience?

Writing style is much like style in anything else. It reveals your personality and your voice as a writer and influences how the reader receives your message. This chapter discusses these basic elements that define style in writing:

- Tone
- Accuracy
- Conciseness
- Language Use
- Sentence Structure
- Passive vs. Active Voice

How Do We Reveal Our Style?

To get a better picture of how we reveal our style as writers, let's look at an analogy. Picture a person walking down the street with hair that is metallic purple, shoulder length, naturally curly, blowing in the wind. How would you describe this person's hairstyle? Perhaps you'd say the style is outgoing, uninhibited, rebellious, attention getting.

Now, picture a second person walking by with hair pulled tightly away from the face into a bun at the nape of the neck. How would you describe this person's hairstyle? Perhaps you'd describe this style as conservative, nonattention getting . . . You get the picture.

Are these two styles appropriate? Of course, these two hairstyles are perfectly fine. But what if I told you the person with the hair that is metallic purple, shoulder length, naturally curly, blowing in the wind is a man in his 80s? How do these details change your assessment of appropriateness? Or do they? Now, let's say this same man is on his way to a black-tie event where he will be surrounded by conservative-thinking (and looking) people. Does your assessment of appropriateness change now?

Like hairstyles, some writing styles are more appropriate for certain situations than others. No writing style will fit all situations. The context within which you are writing determines what writing style is appropriate.

The context includes the medium you're using, the person you're writing to, the topic you're writing about, the purpose of the communication, the organizational climate your intended reader works in, and so on.

For example, if you're writing an e-mail to a friend, you're going to use a different style than if you're writing an e-mail to a work colleague; for your friend, you may use informal language, emoticons, and a plethora of exclamation points. On the formality continuum, you'll use a different style to write an informal e-mail than you'd use to write a formal letter to the company CEO.

Whose Style Is It . . . Really?

Writing rarely happens in a vacuum. Often, in the classroom and workplace, you will find yourself collaborating with others to write a report. When that happens, whose style gets used for the final document? Does the style become a melting pot of all the styles?

A somewhat recent phenomenon on the Web is shared writing. You are most likely familiar with Wikipedia. Wikipedia is a free online encyclopedia that anyone can add to, delete from, or edit. Open source software, which allows others to use and modify it to fit their needs, is also becoming prevalent today. So ownership of content and style is becoming harder to determine today.

Tone

Tone in your writing will determine how your reader responds to your message or even if he or she reads your message . . . which brings us back to purpose and audience. You must know what you're trying to accomplish in the document, and you must know who your audience is.

Have you ever heard the adage "It's not what you say; it's how you say it"? The *how* in how we say something reveals our attitude about what we say. This adage holds true whether we are referring to oral or written communication. In oral communication, the *how* comes through our pitch and volume and body language; in writing, our word choice and the way we put sentences together, for example, reveal the *how*, or our attitude toward the reader and subject matter. We can very easily jinx a deal by using one wrong word or by using the wrong tone!

Let's look at an example: You have a strong opinion about a new procedure that you think should be used in the office; you decide to write to your supervisor to recommend this new procedure. Will your supervisor be more receptive to a *polite requesting* tone or a *negative demanding* tone? Which words are likely to persuade your supervisor to listen to your recommendation: "Based on the data I have presented, I recommend that we implement …" or "Based on the data I have presented, I demand that you implement …"? Do you see how easily we convey our attitude toward something, sometimes without even meaning to?

Let's look at another example: You have a very important idea you want to run by your supervisor. You quickly type your e-mail; you are in a hurry and have no time to run it by a colleague or even step back from it for a few minutes before clicking "Send." Your supervisor replies with a polite "No." You're baffled. You were so certain she would go for it. When you reread your e-mail, you notice that in your haste you had misspelled her name and incorrectly presented a piece of data. The care with which we compose and how well we proofread our documents convey our tone: In this instance, our tone comes across as careless and not caring enough about the idea or the supervisor to even proofread the e-mail.

So, *tone* is the attitude we reveal in our writing about our reader and the subject matter. We convey that tone in several different ways:

- The words we use
- The way we phrase our sentences
- The care with which we edit and proofread our documents
- The accuracy of the information we share
- The professional appearance of our documents
- The amount of effort we put into our research
- The accuracy of our punctuation, spelling, and mechanics

Step back and read what you've written. Ask yourself: If I received this document, how would I react? What tone am I conveying? Am I demeaning the reader in any way? Am I being negative, when I want to be positive? Have I given too much information (implying the reader is dumb) or too little information (implying that I do not have all the facts)?

A FEW EXAMPLES OF TONES FOUND IN WRITING

- Negative
- Excited
- Apologetic
- Formal
- Positive
- Sarcastic
- Compromising
- Uncertain
- Demeaning
- Demanding
- Interested
- Persuasive
- Empowering
- Humorous
- Confused
- Bitter
- Uninterested
- Friendly
- Ironic
- Informal

You may find all of these tones in technical writing, but, obviously, some are more common and appropriate than others. Bottom line: Ask yourself how you want your reader to respond. Answer that question by using the appropriate tone in your writing!

Accuracy

In technical writing, accuracy is critical; it can be the difference between landing a multimillion-dollar contract and losing it. Accuracy can be the difference between users being able to complete a task or not, or it can be the difference between workers safely completing the task or workers ending up in the hospital.

Your document will be only as accurate and complete as the data you gather in your research. Therefore, do meticulous and thorough research! And, report data and information honestly and ethically.

In some work settings, documents must go through a rigorous review process to check for accuracy. With instructions, for example, accuracy is critical.

Conciseness

Technical writing demands conciseness. The cardinal rule in technical communication is get to the point quickly and concisely. Your reader does not want to read words, sentences, paragraphs that could easily be omitted without affecting the meaning of your message, so aim for conciseness. Don't be curt, however. And, do use transitions to make your writing flow smoothly.

THERE IS . . . THERE ARE . . . THERE WILL BE . . .

One easy strategy to help your writing be more concise and more effective is to eliminate, whenever possible, the use of "There is . . . there are . . . there will be . . . there was . . . , etc." to begin sentences. Since the word *there* is not your subject, find the real subject and begin the sentence with that word. Of course, you can't eliminate every "there is . . ." construction in your writing, but when you try, you'll be amazed at how many you can eliminate.

EXAMPLE

- *There are* ten software developers who will attend the meeting.
- Ten software developers will attend the meeting.

Notice how eliminating the *there are* construction makes your sentence more concise and dynamic.

THIS (WHAT?) AND IT

Another strategy that will help you be more concise and effective is eliminating the use of the vague pronouns *this* and *it*. You can easily confuse the reader if *this* or *it* can refer to several words in the previous sentence. Use the word *this* to refer to a specific noun and not to the general idea of a previous sentence or paragraph. Clarify what *it* refers to in a sentence.

EXAMPLES

- This will work out great in the media room.
- This *system* will work out great in the media room.

Notice how adding the word *system* makes it clear to the reader what *this* refers to.

- It is obvious that the report is incomplete.
- Obviously, the report is incomplete.

Notice how starting with the real subject makes the sentence more concise.

Language Use

The language you use when writing must fit your audience and purpose. However, certain rules of thumb apply no matter what the situation. When possible, choose the simple word over the elaborate, "unsmother" your verbs, choose active voice over passive voice, and choose positive language over negative language.

SIMPLE *vs.* ELABORATE

Readers want to understand what they read. So use language that communicates clearly to your reader. You can do that by using simple words. So, if you have a choice, the rule of thumb is to almost always choose the simple word.

Here's where your reader is important again. When writing to technicians, you may want to use the complex or elaborate word . . . if you know your readers will understand it. When writing to managers or to general readers, you will want to steer away from using technical jargon. On the other hand, if you know the manager was a technician before becoming a manager and will understand the technical terminology, then, by all means, use it.

Sometimes *big* words are necessary, especially in highly technical fields when no simple word exists to replace the technical word. If you know your reader will understand the word, you should be safe in using it. But don't forget about your secondary readers, those readers you didn't know would read your document. When it comes to language, use the lowest common denominator to communicate your message.

SMOTHERED VERBS

Another strategy you can use to make your writing style more effective is to "unsmother" the verbs in your writing. Look for those times when you've hidden or "smothered" a perfectly good verb by making it a noun. Table 8.1 shows some examples.

Table 8.1 "Smothered" Verbs

"Smothered" Verbs	**"Unsmothered" Verbs**
Make a decision (e.g., *Will you make your decision tomorrow?*)	Decide (e.g., *Will you decide tomorrow?*)
Arrive at a conclusion	Conclude
Come to a realization	Realize
Offer (make) a suggestion	Suggest
Make a presentation	Present
Take under consideration	Consider
Make a choice	Choose
Make provision for	Provide
Make a quotation	Quote
Is indicative of	Indicates
Make reference to	Refer
Have a need for	Need
Find a use for	Use
Give assistance	Assist
Make an alteration	Alter
Engage in a discussion	Discuss
Engage in a conversation	Converse
Make an acquisition	Acquire

PASSIVE *vs*. ACTIVE VOICE

Passive =
Subject acted upon
Directs action toward subject
- The report was written by Jane. (six words)

Active =
Subject does acting
Directs action toward object
- Jane wrote the report. (four words)

Passive voice is one of the most common style faults in technical writing. The **passive voice** is dull and wimpy, uses 20 to 30 percent more words than active, and is more difficult to read. The **active voice** has more energy, uses fewer words, and is easier to understand. So when possible, use the active voice.

You need to know how passives are formed if you want to avoid them. Passives are formed with:

- Some form of the helping verb *to be*. (The report *was* written by Jane.)
- The past participle of the main verb. (The report was *written* by Jane.)

Often the sentence will have a *by* phrase, but not always. (The report was written *by Jane*.)

Active voice isn't always the best choice, however. Sometimes the passive is preferred. For example, passive voice can be the better choice in these circumstances:

- The doer is unknown or not relevant. ("The door was locked.")
- You need to soften the tone. ("Your check was written for the wrong amount.")
- You are explaining an operation when a person is not actually involved. ("The error message was generated yesterday.")

POSITIVE *vs.* NEGATIVE LANGUAGE

Studies of memory show negative memories are many times easier to recall after several years than pleasant memories. Negatives, whether in words or actions, stay with us. So the language you use in workplace communication can make or break a deal. Using negative language unnecessarily or even unintentionally can lose the company an existing client or a potential one. At the very least, negative language can cause ill will.

NEGATIVE EXAMPLES

- Your service to me failed to meet minimum standards.
- Danny was never unhappy with Emma's work.

SAME SENTENCES PHRASED MORE POSITIVELY

- Your service to me did not meet the high standards I've come to expect from your staff.
- Danny was always happy with Emma's work.

Always choose positive language over negative when possible; however, be ethical in your communication. Never intentionally mislead or manipulate the reader. Think about a coupon for a FREE meal that excludes in the fine print most of the menu items and most of the days of the week.

NEGATIVES TO AVOID IN WORKPLACE WRITING

- Fault (It's your fault the program didn't work.)
- Reject (I am rejecting your offer.)
- Worthless (Your point is worthless.)
- Doubt (I doubt your sincerity.)
- Complain (I have a complaint about your service.)
- Wrong (Your were wrong in your approach to the situation.)
- Careless (Your employee was careless in his attempt to solve the problem.)
- Failed (You failed to meet my needs.)

Sentence Structure

It's important to assess each situation.

To write an effective document, you need to write effective sentences. Of course, everything discussed in this chapter is about writing effective sentences. Using language the reader understands, unsmothering your verbs, using active voice when possible, being concise, using an appropriate tone, being accurate will help ensure you are writing effective sentences. The sentence is the building block of any document. If you are writing confusing, awkward, wordy sentences, you will lose your reader. If the reader has to read a sentence three or four times to figure out your meaning, something's wrong.

PARALLELISM

Placing parallel ideas in parallel structure makes sentences more effective. Parallelism creates a flow in our writing and lets the reader know how ideas relate to one another.

The concept of *parallelism* is closely related to *coordination*. Parallelism specifically addresses items in a series and suggests that because all the items have equal weight, they should all have identical grammatical structure. Thus, you have Abraham Lincoln's famous ending to the Gettysburg Address: ". . . and that government of the people, by the people, for the people shall not perish from the earth." Three things are important about the American government—none is more important than any of the others—and they are rendered in three matching prepositional phrases. The balance of the ideas is reflected in the balance of the grammatical structure.

Good writing practice suggests that you do the same in your technical communication. For instance, in a comparison report to your manager on converting the sales force's cell phone plan from traditional wireless cell phones to smart phones, you decide that a thorough discussion needs an *analysis* of the current practice, a *projection* of the added advantages of adopting smart phones, a *cost analysis* of the conversion and ongoing application, and a *timeline* for implementing your idea. Note that all the words in italics are nouns. They represent the major ideas of your study, the major headings of your report, and by all being in the same grammatical structure, their equality is clear. You may choose to open your report with a summary of your main points, presented in a bullet format:

This report will cover:
- *Current cell phone usage and applications*
- *Projected advantages of converting to smart phones*
- *Cost analysis of conversion and operation*
- *Timeline*

Compare that list to one not as carefully rendered:

This report will cover:
- *Current cell phone usage and applications*
- *Converting to smart phones*

- *Impact on budgets*
- *Can be completed by May 1, 2010*

With a jumbled list, it is harder for the reader to discern the parallel ideas. The rhythm is not there; the balance is missing.

WHICH IS MORE EFFECTIVE?

- *Give me liberty or give me death.* (Patrick Henry's historic speech on March 23, 1775)

Or

- *Give me liberty or kill me.*

- *Ask not what your country can do for you; ask what you can do for your country.* (John F. Kennedy's Inaugural Address, 1960)

Or

- *Ask not what your country can do for you; think about what you might be able to do for your country.*

Chapter Summary

Writing style reveals your personality, and that personality needs to fit the context of your communication, the reader, and your purpose. This chapter discussed how style is revealed in the communication's tone, accuracy, conciseness, language, use of passive vs. active voice, and sentence structure.

End of Chapter Activities

ACTIVITY 1 Editing and Proofreading

Read the following and mark any errors you find.

1. When provoked, Gail has been known to become violet.
2. We want our managers to lie in the communities where our plants are located.
3. The aid altared the report to give the illusion that she was innocent of all charges the counsel had brought against her.
4. Thank you for your letter of April 24th. Let me try to answer each of the questions questions you raised. First, we do not sell the components separately, they only come packaged as a a set.
5. Everyone of the sales representatives have made less calls in the past 6 months then they did in the previous 6-month period.

ACTIVITY 2 Proofreading

1. Read once through the *Ode to the Typographical Error;* don't study it—just read through it.
2. After reading it, write down any proofreading errors you saw.
3. Now go back and reread. Do you see any errors you didn't see the first time?

Paris
in the
the spring

The typographical error is a slippery thing and sly;
You can hunt till you're dizzy, but it'll somehow get by.
Till the forms are on the press, it is strange how still it keeps.
It shrinks down in a corner, and it never stirs or peeps—
That typographical error, too small for human eyes—
Till the ink is on the paper, when it grows to mountain size.
The boss, she stares with horror;
Then she grabs her hair and groans.
The copyreader drops his head upon his hands and moans.
The remainder of the issue may be clean as clean can be,
But the typographical error is the only thing you see.

—Anonymous

ACTIVITY 3 Active and Passive Voice Verbs

Read the following sentences and identify whether they use active or passive verbs.

____________1. Carolyn will lead the committee's discussion.
____________2. The report was written by a committee of well-known experts on the subject.
____________3. Unfortunately, the decision was made without consulting me.
____________4. The department supervisor reports that two employees have questions about the nonsmoking policy.
____________5. Two computers were reported stolen over the weekend by the head of corporate security.
____________6. Danny speaks his mind without regard to the situation.
____________7. The sales staff left the meeting very nervous.

ACTIVITY 4 Parallelism

Correct any faulty parallelism you find in the following sentences.

1. Before leaving the job, make sure you do the following:
   - Put all tools away.
   - Clean up your work area.
   - All windows should be closed.
2. The new software is easy to install, easy to operate, and it is relatively inexpensive.

ACTIVITY 5 "There is/There are," "It," and "This"

Revise the sentences in the following paragraph to eliminate *there is/there are* constructions, wordy or vague *it* constructions, and disconnected *this* constructions.

There are several reasons that we are postponing the company picnic this year. First of all, there is construction being done on the park where we usually hold the picnic, and this will not be completed until mid-August. Second, there are a lot of people in the company who enjoy going to the NASCAR race in August, and it is difficult for them to choose between the race and the picnic. This makes them unhappy. Finally, there are a lot of employees who take vacation in August, and this means that they often miss the picnic. So in the interest of everyone, the company picnic will take place this year in September, and there will be a new date announced for it very soon.

CHAPTER 9

Workplace Research

Overview

Is research something you do only at the university? No, it isn't. Though we would like to have all the information we ever need in our heads at all times and for all situations, that just doesn't happen. We need input from outside sources to help us make informed, evidence-based decisions in the workplace.

Technology gives us access to more knowledge than we know what to do with. Actually, technology has changed the way we think of knowledge. The know-how and know-what are being supplemented with know-where. Because we are exposed to so much knowledge, we must know *where* to find information and *how* to find it quickly.

So you will find yourself doing research in the workplace, but it may look different from the research you do at the university . . . or it may look the same.

In the workplace, you will use research to help you and others make evidence-based decisions. For example, your supervisor tells you he'd like a written report by tomorrow at 3 P.M. on how the Ecosystem Lab is working out for the team. You may go to the Web and check out what others say about the benefits of an ecosystem lab. That's research. You may gather the team and ask them how the lab is working out. That's research. Or you may go to a file you inherited from the person who previously had your job to see if it contains any relevant historical information on the Ecosystem Lab. That's research. And, remember, each organization will have certain acceptable standards or customs when it comes to research. Learn what they are and follow them.

The research you do in the workplace will depend on many things, but as you've gathered by now, your research needs to achieve the specific purpose you want and provide the information the audience or reader wants and needs. And, it must be ethical—that is, it must present information honestly, accurately, and completely. All research involves ethical decisions and legal responsibilities. In library or online research, choosing reliable and credible sources is critical; in primary research, choosing a representative sample to survey or phrasing survey questions objectively can make the difference between ethical research data and unethical research data. In primary research, the rights of respondents must be respected.

Research can be categorized broadly into *primary research* and *secondary research*. This chapter explores several approaches to each type of research.

Purposes of Workplace Research

Companies do external and internal research to understand market trends, to determine competitors' approaches, to make financial decisions, and to establish company policies. They do research prior to and during product development and in mergers and acquisitions. Companies also do research to shorten employee learning curves, to gauge the effectiveness of a change in policy or procedure, to minimize employee resistance to change, and to give employees a sense of being valued and of being heard. Research is a common practice in business and industry.

Primary Research

Primary research is research you do yourself and is an activity you will find yourself involved in often as you make informed decisions on the job. You may not use a formal process, so may not think of it as research, but any time you get feedback from your colleagues or use company files and records to make an informed decision, you are engaged in primary research. So primary research includes such activities as:

- Talking to colleagues
- Looking at company files and records (Company annual reports or documents published by the company fit in this category.)
- Surveying clients
- Interviewing clients
- Conducting focus groups
- Conducting think tanks

Sometimes you need only to talk to colleagues for the information you need. Other times, you'll use a more systematic approach to your research, which takes planning. You want to make sure the data you gather help you make an informed decision or write a report that will help your supervisor make an informed decision. Ask yourself:

1. What do I need to know?
2. What kind of information do I need that I can't find in the library or in an online search?
3. Where can I get this information? Fellow employees? Company files? Published company documents? The general public?
4. Will it be accurate, reliable, and sufficient?

How you answer these questions will determine whether you need quantitative or qualitative data or both. You have many research tools to choose from; however, you should choose the tool that best yields the data you need. For example, if you need or want to know *how many*, questionnaires are an effective and often used tool. However, if you want to know what people think in their own words, interviews or focus groups are effective tools. Be aware that you will need approval from the powers-that-be for most (if not all) research projects in your workplace.

QUANTITATIVE RESEARCH TOOLS

Questionnaires and content analyses are commonly used to gather quantitative data.

Survey by Questionnaire

As mentioned earlier, if you need to know "how many," a questionnaire is a good tool choice. *Quantitative* means capable of being measured. (Note: Questionnaires can also yield qualitative data when open-ended questions are included.) Questionnaires allow respondents to remain anonymous and allow the researcher to engage large numbers of people. By choosing a representative sample group to survey, you can describe something statistically about that group. Whether you can generalize to the larger group depends on how representative your survey group is and how large your sample is. Questionnaires can be distributed in person or via e-mail or the telephone. If you send the questionnaire by e-mail, explain in the e-mail message why you're asking the recipient to complete the questionnaire.

A rule of thumb for designing questionnaires is to keep them as brief as possible to gather the data you need and to make them easy to understand and fill out. See Figure 9.1 for a partial sample questionnaire.

Steps and Tips for Surveying by Questionnaire

You've decided to use a questionnaire to gather data. Here are several steps and tips to consider as you plan your strategy.

1. Choose your survey population carefully. *It must represent the larger group you are researching.*
2. Decide how and when you will distribute your questionnaire. *Will you e-mail, hand deliver, or use postal mail? How much time do you need to tabulate and analyze the data? Based on this information and how long you're giving the participants to respond, you can decide when you need to distribute the questionnaire.*
3. Decide what demographic information (age, gender, etc.) you need from the survey population. *Since it's important not to waste people's time, don't ask questions you don't need answers to. This takes careful planning on your part.*
4. Decide on the question format you'll use: closed-ended or open-ended. *Closed-ended questions (e.g., yes/no, multiple choice, ranking, strongly agree to disagree range, or a combination) limit the responses and are the easiest to tabulate. Open-ended questions require more than one or two word answers and take more time to interpret. They yield qualitative data. That's not to say you should not include open-ended questions; they often result in rich data*

Figure 9.1
Sample Partial Questionnaire

Engineering Student Perceptions of Technical Communication Survey

This survey is part of a research study I'm conducting on engineering student perceptions of technical communication. I'm asking for your personal attitudes! Please be honest; your responses will remain completely anonymous.

Please complete the survey following the directions given below and place in the box provided. Thank you for taking the time to complete this survey and for participating in this study.

Part 1: Perceptions of Technical Communication

Please indicate your agreement or disagreement with each of the following statements by checking the appropriate box.

	Strongly Agree	Agree	Neutral	Disagree	Strongly Disagree
1. I am a good writer.					
2. I am a good oral presenter.					
3. I value good writing skills.					
4. I value good oral communication skills.					
5. Good writing skills are important for my profession.					
6. Good oral communication skills are important for my profession.					
7. This course offers valuable information that I'll be able to use in my other classes.					
8. This course offers valuable information that I'll be able to use after I graduate.					

•••

Figure 9.1 Sample Partial Questionnaire *(continued)*

32. What are your expectations of a technical communication course?

33. Describe to your good friend what technical communication means.

34. Please add any other comments you'd like to make about writing and speaking skills.

Part 2: Background Information

35. Which of the following categories best represents your age?
    _____18–19 years
    _____20 years
    _____21 years
    _____22 years
    _____23–29 years
    _____30–40 years
    _____40+ years

36. What is your gender?
    _____Male ____Female

37. Which of the following represents your year in college?
    _____Sophomore
    _____Junior
    _____Senior

38. What is your major?
    _____Computer Engineering
    _____Electrical Engineering
    _____Mechanical Engineering
    _____Biomedical Engineering
    _____Other (Please specify: ___________)

39. How many writing courses have you taken since starting college?
    _____ None _____ 2 courses
    _____ 1 course _____ More than 2 courses

Thank You!

and unexpected surprises. Refer to questions 32–34 in Figure 9.1 for sample open-ended questions.

5. Draft the survey questions. *Sounds easy, but drafting questions that yield the information you need is the hardest part. Remember, you can't go back and redo the survey (at least not easily or conveniently), so make sure you know what data and demographic information you'll need later. Also make sure you haven't biased the results by the way you worded the questions.*
6. Sequence the questions in a logical order. *Group similar questions together. Get rid of any redundancy. Since people like questionnaires to be quick and easy, consider putting the easiest questions at the beginning and open-ended questions at the end of the questionnaire.*
7. Draft the entire questionnaire.
   a. Create an effective title.
   b. Draft a brief introduction.
      i. Include the questionnaire's purpose.
      ii. Include instructions on how to complete, when to complete, and where to return the questionnaire.
      iii. Let survey participants know how the results will be used and how their privacy will be protected.
      iv. Thank the respondents.
   c. Thank the respondents again!
8. Edit to ensure the questionnaire is complete and easy to follow and the questions are brief, unbiased, and clear.
9. If at all possible, test-run your questionnaire. At the very least, get reader feedback. Revise as needed!
10. Distribute and administer the survey.
11. Tabulate and analyze the results.

Tabulating Primary Data

Once you've collected your data, you will need to analyze and interpret it. In other words, you must explain what the data mean. You can't expect your reader to figure out what all the numbers signify.

An easy way to tabulate questionnaire/survey data is to record the responses to each question by using hash marks on a blank questionnaire or similar form. You'll need to then convert these numbers into percentages.

For example, let's say you've surveyed fifty people on the question *What position do you hold in the company?* Here are the raw data and the data converted to percentages.

3	upper management	(3 divided by 50 = 6%)
7	middle management	(7 divided by 50 = 14%)
15	technician	(15 divided by 50 = 30%)
20	field worker	(20 divided by 50 = 40%)
5	clerical/office worker	(5 divided by 50 = 10%)
0	none of the above (please specify)	

Of course, for online, Web, and e-mail surveys, survey software programs are abundant and will do this work for you. One commonly used survey software program is SurveyMonkey; others you may be familiar with are Zoomerang, QuestionPro, and Polldaddy. These programs are easy to use and take the guesswork out of formatting. Many of these software programs are free for basic surveys.

Analyzing the Data in a Visual Format

Once you have your data tabulated, you can design visuals to represent the data, which will help your readers understand your textual explanation. Let's first convert the data into a visual. A pie chart will illustrate the data nicely. See Figure 9.2. As mentioned earlier, survey software programs will convert the data into charts and tables. Now . . . what do the numbers mean? Remember, you can't throw a figure into your report and expect it to stand alone. You need to explain it. In other words, you need to interpret the data, draw conclusions from the data. What do the data mean? Why are the data significant?

Figure 9.2 illustrates responses to one question on the questionnaire you've administered. Let's assume your questionnaire has twenty questions. Will you want to include in your informal report a visual for every question? Will you want to report specific percentages for every question in your informal report? Most likely not. You will want, however, to summarize the significance of the data and illustrate the most important information in the report. For example, you can write, "A clear majority of employees (70%) falls in the technician and field worker positions." You may also want to include additional visuals and raw data in an appendix.

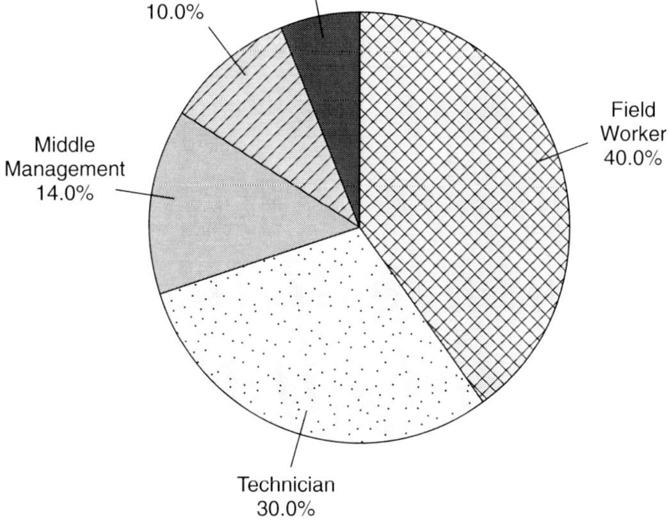

Figure 9.2
Sample Visual, Pie Graph

Source: Questionnaire of XYZ Company, March 20–

- Label, number, & title visuals
- Give source of information

Content Analysis

Another quantitative research tool is the content analysis, which allows you to research changes or patterns of an activity or topic by examining company annual reports, records, minutes of meetings over time, or other kinds of company documents. A content analysis takes careful planning and involves much time and effort.

QUALITATIVE RESEARCH TOOLS

Unlike quantitative data, or numerical information, qualitative data are words. Interviewing is commonly used to gather qualitative data.

Survey by Interviewing Individuals

You may decide the tool you want to use is interviewing. Maybe you want to interview an expert on the topic or several colleagues in a particular department. Interviewing, of course, is more labor intensive than surveying by questionnaire and gives you a wealth of qualitative data. Qualitative data are more difficult and time consuming to analyze than quantitative data. You will want to plan any qualitative study carefully so that you can accomplish your research goal in the least amount of time and with the least amount of trouble for you and the interviewee. The more detailed planning you do, the more likely you'll accomplish this goal.

Setting Up the Interview

When setting up the interview, clearly explain who you are, what kind of information you're looking for, and why you chose this particular person to interview. Remember to explain how you will use the information and how you will protect the participant's privacy. If you're interviewing more than one person on the same topic, you'll need to ask the exact same questions and in exactly the same way. So, you'll want to develop an interview guide to help you keep on track.

Conducting the Interview

Here are some guidelines to follow when interviewing:

- Be courteous and take only the time you need to get specific questions answered. *If you gave a specific amount of time the interview would take, honor that.*
- Follow the interview guide you prepared ahead of time; however, don't pass up a chance to ask a follow-up question that naturally arises during the interview.
- Let the interviewee do most of the talking!
- If the interviewee gets off topic, don't be afraid to bring her or him back to the questions you need answered.
- Take only notes that you really need to take during the interview.
- Audio or video record the interview, if possible; however, you must get permission beforehand. Understand, also, that some people are intimidated by being recorded and may not give you the quality interview

they might otherwise. Another possible disadvantage is that your equipment may not work properly. Even when recording, relying solely on the recording is not a good idea. For example, you may think your equipment is working properly when it is not. Taking some notes on the very important information is always a good idea.

- Thank the interviewee at the end of the interview.
- If you have audio or video recorded the interview, check it immediately following the interview to see if you have a quality recording. If not or if you did not record, look at your notes and fill in any gaps. Also take the time to write out the important points in complete sentences. Note what parts were most important. If you have direct quotes you may want to use, write them down now and use quotation marks to remind yourself they are direct quotes.

Following Up after the Interview

It is common courtesy to run anything you quote directly in your report by the interviewee. If the topic is sensitive, you may want to let the interviewee review what you have written, even if you are not quoting the person directly.

Remember to send the interviewee a follow-up thank you note.

Survey by Interviewing Groups

Focus groups and think tanks are means of interviewing multiple people at the same time. These two methods often result in complex and in-depth data nearly impossible to gather in one-on-one interviews. Audio or video recording the group session is the best way to retrieve all the information shared during the session.

VIDEO RECORDING

Technology is making it increasingly easier to video record one-on-one interviews, focus groups, and think tanks. For example, camcorders are small enough to fit into a pocket, easy to operate, and relatively inexpensive, and produce quality recordings.

Secondary Research

Secondary research, unlike primary research, is using work completed by others. By using work completed by others, you can support your own ideas. If you show that others who are credible support your ideas, your ideas become more credible and thus more persuasive.

Historically, of course, we think of secondary research as library research—visiting the physical building which houses hardbound books, magazines, journals, photographs, microfilm, audio and videotapes . . . you get the idea. However, the face of the library has changed dramatically with the growth of technology. Libraries are now also virtual storage spaces for e-books and electronic sources of all kinds that can be accessed online or virtually through electronic databases.

This section focuses on searching library electronic sources and the World Wide Web for needed information.

Common Electronic Databases Found in Libraries

- The library's online catalog
- Bibliographic databases (citations and abstracts)
- Full-text databases (complete text of a work)
- Aggregated databases (collections of different databases, such as *EBSCOhost*)

Basic Ways to Search Electronic Databases

- By author's name
- By title of the work
- By the subject of the work
- By words or keywords

LIBRARY ELECTRONIC SOURCES

Retrieving data from a library electronic database is an excellent way to begin your research if you choose not to physically go to a library. Technology makes retrieving data fast and easy; however, finding the data you want and need can be a daunting task. If you don't use an effective search process, you can waste many hours of valuable time. Using an effective search process is key to finding data you need and want.

Typical ways to search an electronic database are by the author's name, by the title of the work, by the subject of the work, and by words or keywords.

Searching by **author's name** or **title of work** means you know at least part of the author's name and title. Knowing the entire name or title will give you a more precise search result; however, knowing only part of it will usually result in success as well.

A **subject search** will give you a more precise result than a keywords search; however, you need to know the exact words as they appear in the subject field; you may know these subject headings as *descriptors*. For example, if you're looking for information on *streetcars*, you will need to type in *cable cars*.

Searching by **keywords** means the words or phrases can be found anywhere in the document. You will often need to refine your search. The most common tools to help you refine your search include Boolean operators, quotation marks, truncation, and parenthesis. Most search engines recognize these tools; however, some may not. Some search engines are also case sensitive. Check the help section of the search engine for clarification.

Boolean Operators

Boolean operators are *and*, *or*, and *not*. By adding an *and* between two keywords, your search will yield sources that contain both keywords. Adding a *not* between two keywords results in sources that contain the first keyword but not the second. Adding an *or* between keywords results in sources that contain either keyword.

Quotation Marks

Placing quotation marks around a phrase ("technical communication") results in sources that contain all the words in the phrase and in the same order.

Truncation

Truncation is placing an asterisk (*) at the end of the root word, which results in all the variations of the root word. For example, if you truncate the word notif*, the search engine will find notify, notification, notifying, etc.

Parentheses

Parentheses allow you to chunk information, which search engines will execute first before other keywords. For example, with (*attitude* OR *perception*) AND *success*, the search engine will process *attitude OR perception* first.

INTERNET/WORLD WIDE WEB

To do research today, one of the first places many of us think about going to is the Internet. However, as more and more knowledge is stored on the Internet, this medium may become difficult to maneuver in the near future.

The two means used most often to find information on the Web are search engines and directories. Google (http://www.google.com/), AltaVista™ (http://www.altavista.com/), and Ask (formerly Ask Jeeves) (http://www.ask.com/) are commonly used search engines. Unlike search engines that find and list Web sites based on keywords, directories organize the Web by subject and then by categories. Google is also a directory. Another common directory is Yahoo (http://www.yahoo.com/).

Meta search engines search the search engines, so if you know clearly what information you're after, a meta search is effective. Common meta search engines include Dogpile (http://www.dogpile.com/), MetaCrawler (http://www.metacrawler.com/index.html), and Ixquick (http://www.ixquick.com/).

As we know, anyone can publish anything on the Internet, so the information is fluid, unlike the solid nature of materials found in the library. Being unregulated, there's no guarantee the information you find on the Web is credible or reliable, making it very important to do your due diligence.

Using the Internet definitely has its pluses, however. Ease and speed of use are huge in today's chaotic world. Because of this ease and speed of use, many people use the Internet indiscriminately. Don't! You need to make sure the sources you find on the Internet are credible.

Domain types tell you a great deal about whether a source is credible and reliable. Here are a few common domain types:

- .com (commercial)
- .edu (U.S. educational institution)
- .org (organization—usually not-for-profit)
- .gov (U.S. government)
- .mil (U.S. military)
- .net (networking)
- .biz (business in the U.S.)
- .info (informational sites)

Some domains we can assume to be credible: .edu, .gov, .mil. Other domains may have credible and not-so-credible sources. For example, .org, usually reserved for not-for-profits, is also used by professional organizations and political parties and anyone who wants to use it. The .com domain may have credible and not so credible sources also, now that anyone can register a .com

Evaluating Web Sources

1. Who is the author? What are his credentials, qualifications?
2. Who is the publisher? Is the publisher reputable?
3. Is the information accurate? Do you know the source of the information?
4. Is the information objective? Do you notice a bias in either the information given or the way the information is given?
5. Is the information current?

domain. The bottom line: You must evaluate the credibility and reliability of most sources found on the Web.

TRADE PUBLICATIONS

Trade publications are an excellent source of information in the workplace. Trade publications are specifically geared to a particular industry. Articles use technical language and so are often dense with information that appeals only to those in the industry. Even the advertising focuses on the specific industry. See Table 9.1 for a brief comparison of trade journals, scholarly journals, and popular magazines.

Table 9.1 Trade Journals, Scholarly Journals, and Popular Magazines Compared

	Trade Journals	**Scholarly Journals**	**Popular Magazines**
Written for	Workers in a specific industry	Professors and students	General public
Written by	Professional writers and industry experts	Scholars	Professional writers
Content focused on	Industry trends and news	Research, theory, and scholarship	General topics of interest to the public
Examples:	• *CIO* • *Public Management* • *Advertising Age*	• *Journal of Applied Psychology* • *Science* • *Journal of Marketing Research*	• *Bon Appetit* • *Time* • *Psychology Today*

TIPS FOR SEARCHING

Here are tips for electronic searching:

1. Determine specifically what you're trying to accomplish in your research.
2. Write down your research question(s) and keep visible to ensure you stay focused.
3. Determine specifically what information you're looking for.
4. Write down a list of databases relevant to your research topic.
5. Write down a list of keywords that are relevant to your research topic.
6. Review the techniques for combining keywords ("AND," parentheses, quotation marks) and truncating keywords.
7. If you come up with too few hits, make sure your keywords include the "right" words.
    a. Use the thesaurus feature of the database to find a list of subject headings.
    b. Use the Boolean operator "OR" to broaden your search

8. If you come up with too many hits, narrow your search.
   a. Look at the database index to find more precise search terms.
   b. Use the Boolean operators "AND" and "NOT" to narrow your search.
   c. Limit your search by country.
   d. Narrow your topic and begin the process over.
9. Note any source you find that is just what you're looking for and identify the subject terms listed before the abstract.
   a. Add these terms to your search terms.
   b. If you can find the full text of the article that is just what you're looking for, look at the references page to find sources the author(s) used. You may want to use these sources in your own research.
10. Determine if your sources are credible, reliable, and current.

Reference librarians are wonderful resources; talk to one!

Plagiarizing

When using other sources in your written document, you must know how *not* to plagiarize and how to summarize and paraphrase.

You already know to use quotation marks to indicate you are quoting the writer's exact words and punctuation. Remember, though, in addition to using quotation marks, you must also document the source you've used. You must do both when quoting exact words: Document the source and use quotation marks. If either is missing, you have plagiarized. Plagiarism in the academic environment is considered cheating! In the workplace, it may be considered criminal.

No one, however, wants to look at a document with quotation marks around every sentence. Therefore, you will find yourself summarizing and paraphrasing a great deal of the material you use from other sources.

Summarizing and Paraphrasing

To summarize or paraphrase, you must understand the material, which means you must read the material carefully and often more than once.

A summary is usually significantly shorter than the original. When summarizing information from another source, you'll include only the major points. You'll use your own words with no interpretation. Remember, however, you must document your source, even though you're using your own words.

A paraphrase is not a summary. Unlike a summary, a paraphrase will contain more supporting details than the summary and so may not be significantly shorter than the original. When paraphrasing, you restate the material in your own words. And again, you must document your source even though you're using your own words.

Most likely, you will use all three—quotation, summary, and paraphrase—when writing a report based on research material. See Figure 9.3 for a brief comparison of the three terms (paraphrase, summary, and plagiarism) from Purdue University's Online Writing Lab (OWL).

Figure 9.3 Comparison of Paraphrase, Summary, and Plagiarism

Source: Purdue University Online Writing Lab. Retrieved May 21, 2008, from http://owl.english.purdue.edu/OWL/resource/563/02 (Note: Documentation is in MLA style.)

A Comparison

The original passage

Students frequently overuse direct quotation in taking notes, and as a result, they overuse quotations in the final [research] paper. Probably only about 10% of your final manuscript should appear as directly quoted matter. Therefore, you should strive to limit the amount of exact transcribing of source materials while taking notes. Lester, James D. *Writing Research Papers*. 2nd ed. (1976): 46–47.

A legitimate paraphrase

In research papers, students often quote excessively, failing to keep quoted material down to a desirable level. Since the problem usually originates during note taking, it is essential to minimize the material recorded verbatim (Lester 46–47).

An acceptable summary

Students should take just a few notes in direct quotation from sources to help minimize the amount of quoted material in a research paper (Lester 46–47).

A plagiarized version

Students often use too many direct quotations when they take notes, resulting in too many of them in the final research paper. In fact, probably only about 10% of the final copy should consist of directly quoted material. So it is important to limit the amount of source material copied while taking notes.

Documentation Style

Documentation is an extremely important part of research. Proper documentation serves to give credit to the original writer and keeps you honest. It also allows the reader to locate the source and to replicate your research if necessary.

Many publishing companies and professional societies and organizations publish their own manuals of documentation style. Here are only a few of them:

- *Publication Manual of the American Psychological Association* (APA) (http://apastyle.apa.org/)
- *Chicago Manual of Style* (http://www.chicagomanualofstyle.org/tools_citationguide.html)
- *MLA Style Manual for Writers of Research Papers* (Modern Language Association) (http://www.mla.org/)
- *IEEE Editorial Style Manual* (Institute of Electrical & Electronics Engineers) (http://www.ieee.org/portal/cms_docs_iportals/iportals/publications/authors/transjnl/stylemanual.pdf)
- *CBE Style Manual for Authors, Editors, & Publishers* (Council of Biology Editors) (http://www.councilscienceeditors.org/) This style manual is for all the sciences, not just biology.
- *ACS Style Guide: A Manual for Authors and Editors* (American Chemical Society) (http://pubs.acs.org/index.html)
- *A Manual for Authors of Mathematical Papers* (American Mathematical Society) (http://www.ams.org/journals/)

Please refer to a style manual when trying to determine how to document your sources. OWLs (Online Writing Labs) are also good resources to use: The Purdue University Online Writing Lab is an excellent place to start (http://owl.english.purdue.edu/). In addition, software is available to make documentation much easier.

Chapter Summary

The importance of understanding how to conduct effective and ethical workplace research cannot be overstated. Although research is not synonymous with academics, many of the same principles apply, including the important one of acknowledging the source of the information. This chapter has discussed basic tips on conducting primary and secondary research.

End of Chapter Activities

ACTIVITY 1 Drawing Conclusions from Data

Pair and complete this in-class activity. Each person must be willing to share an item that contains data—a wallet or a backpack, for example.

Step 1: Researching

List what you find in your partner's wallet, putting items into groups or categories as much as possible. Example—Sal's wallet has:

1. Driver's license, a club membership, school ID
2. Pictures of 6 cousins and 3 ex-girlfriends, phone number lists, a joke, waterpark expired coupon, several worn movie ticket stubs
3. One blank check, 3 credit cards, 25 credit card receipts, no money

Now make a similar list from the data you've gathered from your partner's wallet.

Step 2: Drawing Inferences from Your Research (making meaning from your data)

Tell what each list of details from Step 1 says about the wallet's owner. Example:

1. Sal likes women, lots of women, and pursues his pleasures.
2. Sal is beginning to enter society, to get established.
3. Sal carefully watches his finances, but women and movies probably keep him low on ready cash.

Now make similar inferences about your lists from Step 1.

Step 3: Summing Up

Make a summation statement from your inferences and data. That is, give an overview of what all the details *taken together* say about the wallet's owner. *Note:* You will have to make choices here about which perspectives to emphasize. Example:

> *Sal is a carefree guy who pursues his pleasures and carefully manages his money to be able to do so.*

Write a similar summation statement of your partner based on your research.

ACTIVITY 2 What You Learned from the Articles—Group In-Class Analysis/Summary Activity

After reading the articles provided by the instructor, form groups of three to five people. Groups do the following in the computer lab:

- Designate a recorder.
- One person in the group shares his/her list of main points gleaned from the assigned article. Recorder types these points (or simply uses each person's e-copy), listing each point on a separate line.
- As each group member shares her/his list of main points, the recorder puts a mark by the ones already on the list to indicate the repetition of that point and adds any new ones at the bottom of the list. Recorder—don't forget to add your own list of ideas.
- Now look at the complete list of main ideas. What ideas fit together? Put these ideas into their own groupings.
- Look at each grouping of ideas and decide the main point they all refer to. Use these main points as headings.

This activity can be used with an individual article/source or with several articles/sources.

ACTIVITY 3 Summarizing

Pair with a classmate and summarize the following piece. Be prepared to share your summary with the entire class.

Congressional Cemetery: 200th Anniversary

Currently celebrating its two-hundredth anniversary, Congressional Cemetery is located on Capitol Hill in Washington, D.C. A privately-run cemetery, it bears the name Congressional because many members of Congress from the nineteenth century are interred there. Beginning in 1807, with the death of Connecticut Senator Uriah Tracy, eighteen senators and forty-two representatives have been buried in this cemetery. Congress also initiated the practice of erecting monuments, "cenotaphs," in memory of all its members who died in office, whether buried in the cemetery or not. Over time, improvements in transportation allowed deceased members to be removed to their home states for burial and in 1877 the practice of erecting cenotaphs was discontinued (although a century later one more was added in memory of two members of the House lost in an Alaskan plane crash).

Along with senators and representatives, the cemetery holds many other individuals associated with the Congress, among them Dr. William Thornton, the designer of the Capitol building; Samuel Otis, the first Secretary of the Senate; Vice President Elbridge Gerry, who died on his way to preside over the Senate; Isaac Bassett, one of the first Senate pages and later a longtime doorkeeper of the Senate; Joseph Gales and William Seaton, early newspapermen who recorded the debates in the Senate and House; Anne Royall, one of the first women journalists to cover Congress; Mathew Brady, who photographed Civil War-era legislators; and several architects, carpenters, artists, stone masons, and clerks who contributed to the operations of the early Congress, as well as others associated with the federal government.

The cemetery is open to the public. The gates close everyday at dusk.

Source: http://www.senate.gov/general/Features/Congressional_Cemetery.htm

ACTIVITY 4 Paraphrasing

Pair with a classmate and paraphrase the piece in Activity 3. Be prepared to share your work with the entire class.

APPENDIX A

Resources

Figure 1 Communication Strategy Worksheet, Comparison Report

Communication Strategy Worksheet

COMPARISON REPORT

Instructions: Complete this Communication Strategy Worksheet for your Comparison Report. Visualize your reader and answer the prompts/questions thoroughly. The more thorough you are, the easier it will be for you to write the report.

1. What is the name of the person to whom you are writing this report?

2. What is your primary reader's relationship (within company) to you (boss, colleague, etc.)?

3. What is your primary reader's job title in the company?

4. Why will your reader be interested in your report?

5. In addition to your primary reader, who else in the company may need to read the report?

6. What does your reader already know about the topic? Need to know about the topic?

7. What reaction do you anticipate from the reader when he/she reads your report? Will the reader classify the situation as high importance, moderate importance, or low importance?

8. Is your reader a manager, an expert, a technician, or a general reader?

9. What three or four items (services, vendors, products, etc.) will you compare?

10. What three or four criteria will you use to compare the items?

11. How and where will you collect the information you will use in your comparison (e.g., will you use a paper survey, will you talk with colleagues, will you review company documents, will you review the company Web site)?

12. Do you plan to make a recommendation?

13. What visuals or attachments might you include in this report?

Comparison Report

COMPARISON LOG

Figure 2
Comparison Log, Comparison Report.

The Comparison Log is an excellent heuristic to use when comparing items. Use one or something similar for your Comparison Report.

The log is easy to use. Here is how it works.

1. Make a table to fit the number of items you are comparing and the number of criteria you are using to judge the items.
2. List the items (products, services, vendors, parts, etc.) you are comparing in the top header row. (Alternatively, you can list them in the far left column.)
3. Think about the needs of your reader(s). What criteria will you use to judge the items? Place these criteria in the far left column. (Alternatively, you can list them in the top header row.)
4. Include these three items:
   - **IDEAL** (your list of criteria)
   - **REAL** (what the product/service is really like)
   - **JUDGMENT** (the difference between the ideal and real) (Realistically, this item can be omitted, depending on the complexity of your comparison.)
5. Write your **OVERALL JUDGMENT** at the bottom of your log.

Here's a simple example with only the first criterion identified.

IDEAL CRITERIA	Widget 1	Widget 2	Widget 3	Widget 4	JUDGMENT			
					W1	W2	W3	W4
Must cost less than $10.00	$9.99	$11.99	$8.00	$10.80	+	–	+	–
Criterion 2								
Criterion 3								
Criterion 4								
Criterion 5								
Criterion 6								

Overall Judgment: Considering all the criteria, Widget 3 is the one our company should buy because it (reason 1), (reason 2), and (reason 3).

Figure 3 Design Worksheet, Comparison Report

Design Worksheet
COMPARISON REPORT

Instructions: Complete the following prompts thoroughly for your Comparison Report.

Beginning
Introduces the subject, provides brief background information, tells the purpose, and prepares the reader for what follows in the report.

Orienting sentence:
Background information:
Purpose statement:

Middle
Presents the details and evidence supporting your conclusions (longest part of the report).

Situation:
Background details (e.g., previous comparisons completed):
Options compared:
Criteria used:
How researched:
Details about options:
Conclusion(s) about options:
Recommendation (if appropriate):
Visuals or attachments to include:

End
Wraps up the report, and ends on a positive note.

Situation (brief):
Major conclusion (brief):
Major recommendation (if appropriate, brief):
Sense of cooperation:

Peer Review Worksheet

Figure 4
Peer Review Worksheet, Comparison Report

COMPARISON REPORT

Instructions: Read your classmate's report, making comments on the draft. Then complete this worksheet completely and thoroughly. Pretend you are the person receiving the report.

Section A: Read the *Beginning* section and answer these questions.

1. If in memo format, does the Subject Line give you a clear and specific idea of the subject of the report? Explain.
2. What is the orienting sentence(s) that gives you the reason you are reading the report?
3. Does the report have a purpose statement? What is it?

Section B: Read the *Middle* section and answer these questions.

1. Does the writer give you adequate background information to understand the purpose of the comparison?
2. Does the writer state the options compared and the criteria used? What are they?
3. Are the criteria reasonable? Explain.
4. Are the criteria thoroughly investigated? Explain.
5. What sources does the writer use to research the options?
6. What are the results of the writer's comparison?
7. Does the recommendation (if there is one) follow naturally from the discussion? Explain.
8. What do you still need to know to fully understand why the writer is comparing these options or making this recommendation (if there is one)?

Section C: Read the *End* section of this report and answer these questions.

1. Does the writer bring the report to a logical close? Explain.
2. Does the writer end with a positive sense of cooperation (offer to provide additional information or meet with the reader)? Explain.
3. Do you have enough information to make an informed decision? Explain.

Section D: Grammar/Spelling/Punctuation

1. Underline all sentences that begin with "There" or "It."
2. Highlight all sentences that are awkward, wordy, or cause you to wonder what they mean.
3. Is the report formatted effectively? Explain.

Section E: Comments

What else can you share with the writer to make this report even more effective?

Figure 5 Communication Strategy Worksheet, Problem Analysis or Problem/Solution Report

Communication Strategy Worksheet

PROBLEM ANALYSIS OR PROBLEM/ SOLUTION REPORT

Instructions: Complete this Communication Strategy Worksheet for your Problem Analysis or Problem/Solution Report. Visualize your reader and answer the prompts/questions thoroughly. The more thorough you are, the easier it will be for you to write the report.

1. What is the name of the person to whom you are writing this report?
2. What is your primary reader's relationship (within company) to you (boss, colleague, etc.)?
3. What is your primary reader's job title in the company?
4. Why will your reader be interested in your report?
5. In addition to your primary reader, who else in the company may need to read the report?
6. What does your reader already know about the topic? Need to know about the topic?
7. What reaction do you anticipate from the reader when he/she reads your report? Will the reader classify the problem as high importance, moderate importance, or low importance?
8. Is your reader a manager, an expert, a technician, or a general reader?
9. Draw an organization chart of your company. Place yourself and the person you are writing the report to on it. On the organization chart, circle or highlight the levels affected by the problem.
10. What problem are you analyzing?
11. What are the short- and long-term effects of the problem? What effect has the problem caused so far? What happens if the problem is not resolved?
12. What do you think is causing the problem?
13. How and where will you find the information you need for your report?
14. What visuals or attachments might you include in this report?

Design Worksheet

PROBLEM ANALYSIS OR PROBLEM/ SOLUTION REPORT

Figure 6
Design Worksheet, Problem Analysis or Problem/Solution Report

Instructions: Complete the following prompts thoroughly for your Problem Analysis or Problem/Solution Report.

Beginning
Introduces the subject, provides brief background information, tells the purpose, and prepares the reader for what follows in the report.

Orienting & background statements (seriousness & scope, whom problem affects):
Purposestatement:

Middle
Presents the details and evidence supporting your conclusions (longest part of the report).

Problem situation:
Background details:
Previous investigations of problem:
Whom problem affects:
Short- and long-term effects of problem:
Cost of problem:
Causes of problem:
How researched:
Conclusion(s):
Solution (if appropriate):
Visuals or attachments to include:

End
Wraps up the report, and ends on a positive note.

Seriousness of problem situation (brief):
Major conclusion (brief):
Major recommendation (if appropriate, brief):
Sense of cooperation:

Figure 7
Peer Review Worksheet, Problem Analysis or Problem/ Solution Report

Peer Review Worksheet

PROBLEM ANALYSIS OR PROBLEM/ SOLUTION REPORT

Instructions: Read your classmate's report, making comments on the draft. Then complete this worksheet thoroughly. Pretend you are the person receiving the report.

Section A: Read the *Beginning* section and answer these questions.

1. If in memo format, does the Subject Line give you a clear and specific idea of the subject of the report? Explain.
2. What is the orienting sentence(s) that gives you the reason you are reading the report?
3. Does the report have a purpose statement? What is it?
4. What is the problem?

Section B: Read the *Middle* section and answer these questions.

1. Does the writer state the problem? What makes the problem significant?
2. How widespread is the problem (whom/what does it affect)?
3. What are the effects of the problem on the organization, the staff, etc.?
4. What are the contributing factors to (causes of) the problem?
5. Has anyone tried to solve the problem before? Explain.
6. How did the writer investigate the problem?
7. What conclusions does the writer make about the problem based on the investigation?
8. What solution (if there is one) does the writer propose?
9. Does the *Middle* section of the report provide enough information and detail to convince you that the writer has done his/her research well? Explain.

Section C: Read the *End* section of this report and answer these questions.

1. Does the writer bring the report to a logical close? Explain.
2. Does the writer summarize the major conclusions about the problem and restate the solution (if there is one)? Explain.
3. Does the writer end with a positive sense of cooperation (offer to provide additional information or meet with the reader)? Explain.
4. What do you still need to know to feel this report is complete?

Section D: Grammar/Spelling/Punctuation

1. Underline all sentences that begin with "There" or "It."
2. Highlight all sentences that are awkward, wordy, or cause you to wonder what they mean.
3. Is the report formatted effectively? Explain.

Section E: Comments

What else can you share with the writer to make this report even more effective?

Communication Strategy Worksheet

RECOMMENDATION REPORT

Figure 8 Communication Strategy Worksheet, Recommendation Report

Instructions: Complete this Communication Strategy Worksheet for your Recommendation Report. Visualize your reader and answer the prompts/questions thoroughly. The more thorough you are, the easier it will be for you to write the report.

1. What is the name of the person to whom you are writing this report?

2. What is your primary reader's relationship (within company) to you (boss, colleague, etc.)?

3. What is your primary reader's job title in the company?

4. Why will your reader be interested in your report?

5. In addition to your primary reader, who else in the company may need to read the report?

6. What does your reader already know about the topic? Need to know about the topic?

7. What reaction do you anticipate from the reader when he/she reads your report? Will the reader classify the situation as high importance, moderate importance, or low importance?

8. Is your reader a manager, an expert, a technician, or a general reader?

9. What do you think you'll recommend?

10. What are the benefits to the company, department, and employees?

11. What are the possible drawbacks to your recommendation? How will you minimize them?

12. What kind of timeline and costs might be associated with your recommendation?

13. How and where will you collect the information you will need for the recommendation report?

14. What questions might your reader want you to answer in the report?

15. What visuals or attachments might you include in this report?

Figure 9
Design Worksheet, Recommendation Report

Design Worksheet

RECOMMENDATION REPORT

Instructions: Complete the following prompts thoroughly for your Recommendation Report.

Beginning
Introduces the subject, provides brief background information, tells the purpose, and prepares the reader for what follows in the report.

Orienting sentence:
Background or situation:
Purpose statement:
Recommendation (if appropriate):

Middle
Presents the details and evidence supporting your conclusions (longest part of the report).

Situation:
Background details:
Previous investigations:
Whom situation affects:
Short- & long-term effects:
How researched:
Recommendation:
How recommendation benefits company, department, employees:
Possible drawbacks of recommendation and how to minimize:
Timeline for implementation:
Cost of recommendation:
Visuals or attachments to include:

End
Wraps up the report, and ends on a positive note.

Situation (brief):
Recommendation (brief):
Major benefit to the company:
Next step (if appropriate):
Sense of cooperation:

Peer Review Worksheet

Figure 10 Peer Review Worksheet, Recommendation Report

RECOMMENDATION REPORT

Instructions: Read your classmate's report, making comments on the draft. Then complete this worksheet thoroughly. Pretend you are the person receiving the report.

Section A: Read the *Beginning* section and answer these questions.

1. If in memo format, does the Subject Line give you a clear and specific idea of the subject of the report? Explain.
2. What is the orienting sentence(s) that gives you the reason you are reading the report?
3. Does the report have a purpose statement? What is it?
4. Does the report have a recommendation in the introduction? What is it?
5. What do you still need to know to understand why you are receiving the report and what you are supposed to do with it?

Section B: Read the *Middle* section and answer these questions.

1. Does the writer state the situation and its significance? What effect does the situation have on the company, other employees, or the reader?
2. How does the writer investigate/research the situation?
3. Does the writer mention any criteria he/she used to determine the recommendation?
4. What conclusions does the writer draw from the investigation/research?
5. Does the writer give a recommendation in the *Middle* section? If yes, does it flow naturally from the data supplied? Explain.
6. Are the benefits of the recommendation clearly stated?
7. Does the writer mention any drawbacks? What are they? How are they minimized?
8. Does the writer mention a timeline or costs associated with the recommendation? If no, should he/she?
9. Does the *Middle* section provide you with enough data to convince you that the writer has done his/her research well and has reached a reasonable recommendation? Explain.
10. Are visuals included in the report? If yes, does the writer refer to them and explain them in the text of the report? Are they effective? Explain.
11. What do you still need to know to fully "buy in" to the recommendation?

Figure 10 Peer Review Worksheet, Recommendation Report *(continued)*

Peer Review Worksheet

RECOMMENDATION REPORT

Section C: Read the *End* section and answer these questions.

1. Does the writer bring the report to a logical close? Explain.
2. Do you find a summary statement, a recommendation, or a final conclusive statement?
3. Does the writer state or restate the recommendation?
4. Does the writer tie everything together effectively and leave you feeling comfortable with the information provided? Explain.
5. Does the writer end with a positive sense of cooperation (that is, offer to provide additional information or meet with the reader)? Explain.
6. As the person receiving the report, would you act on the recommendation? Why or why not?

Section D: Grammar/Spelling/Punctuation

1. Underline all sentences that begin with "There" or "It."
2. Highlight all sentences that are awkward, wordy, or cause you to wonder what they mean.
3. Is the report formatted effectively? Explain.

Section E: Comments

What else can you share with the writer to make this report even more effective?

Communication Strategy Worksheet

Figure 11
Communication Strategy Worksheet, Lab Report

LAB REPORT

Instructions: Complete this Communication Strategy Worksheet for your Lab Report. Visualize your reader and answer the prompts/questions thoroughly. The more thorough you are, the easier it will be for you to write the report.

1. What is the name of the person to whom you are writing this report?
2. What is your primary reader's relationship (within company) to you (boss, colleague, etc.)?
3. What is your primary reader's job title in the company?
4. Is your reader a manager, an expert, a technician, or a general reader?
5. In addition to your primary reader, who else in the company may need to read the report?
6. What is the purpose, objective, or hypothesis of your lab work?
7. What background information do you need to include in your lab report? How much does your reader already know about the topic? Need to know about the topic?
8. What lab materials or equipment will you need to complete your lab work?
9. What procedures or methods will you use in your lab work?
10. What do you project the results of your lab work will be?
11. What is the significance of your lab results from a practical viewpoint? From your hypothetical (or real) industry client's viewpoint?
12. What visuals or attachments might you include in this report?

Figure 12
Design Worksheet, Lab Report

Design Worksheet
LAB REPORT

Instructions: Complete the following prompts thoroughly for your Lab Report.

Beginning
Introduces the subject, provides brief background information, tells the purpose, and prepares the reader for what follows in the report.

Purpose statement:
Purpose of lab work (brief):
Background information:
Procedure used (brief):
Major results of lab work (brief):

Middle
Presents the details and evidence supporting your conclusions (longest part of the report).

Objective of lab work:
Expected results:
Equipment used in lab work:
Methods & procedures used:
Unusual problems encountered:
Detailed results of lab work:
Significance of results:
Visuals or attachments to include:

End
Wraps up the report, and ends on a positive note.

Major results/conclusions of lab work (brief):
Implications for further work needed:
Sense of cooperation:

Peer Review Worksheet

Figure 13 Peer Review Worksheet, Lab Report

LAB REPORT

Instructions: Read your classmate's report, making comments on the draft. Then complete this worksheet thoroughly. Pretend you are the person receiving the report.

Section A: Read the *Beginning* section and answer these questions.

1. Does the report have a purpose statement? What is it?
2. What lab work does the writer perform?
3. What background information does the writer share?
4. What procedures does the writer use to perform the lab work?
5. What are the results of the lab work?

Section B: Read the *Middle* section and answer these questions.

1. What is the object of the lab work?
2. What results did the writer expect?
3. What equipment is used in the lab work?
4. What procedures are used?
5. What unusual problems are encountered in the lab work, if any?
6. What are the detailed results of the lab work?
7. What visuals or attachments are included in the report? Does the writer explain and refer to them in the report?
8. Does the *Middle* section provide enough data to convince you the writer has done the lab work well and has reached a reasonable conclusion? Explain.

Section C: Read the *End* section and answer these questions.

1. Does the report have a conclusion? What is it?
2. Does the conclusion include further work needed? Explain.
3. Does the writer tie everything together appropriately and leave you feeling comfortable with the information provided? Explain.
4. Does the writer offer to provide additional information or to meet with the reader?
5. What do you still need to know to feel this report is complete?

Section D: Grammar/Spelling/Punctuation

1. Underline all sentences that begin with "There" or "It."
2. Highlight all sentences that are awkward, wordy, or cause you to wonder what they mean.
3. Is the report formatted effectively? Explain.

Section E: Comments

What else can you share with the writer to make this report even more effective?

Figure 14 Communication Strategy Worksheet, Progress Report

Communication Strategy Worksheet

PROGRESS REPORT

Instructions: Complete this Communication Strategy Worksheet for your Progress Report. Visualize your reader and answer the prompts/questions thoroughly. The more thorough you are, the easier it will be for you to write the report.

1. What is the name of the person to whom you are writing this report?
2. What is your primary reader's relationship (within company) to you (boss, colleague, etc.)?
3. What is your primary reader's job title in the company?
4. Why will your reader be interested in your report?
5. In addition to your primary reader, who else in the company may need to read the report?
6. What does your reader already know about the topic? Need to know about the topic?
7. What reaction do you anticipate from the reader when he/she reads your report? Will the reader classify the situation as high importance, moderate importance, or low importance?
8. Is your reader a manager, an expert, a technician, or a general reader?
9. What is the project you are reporting on? What are the objectives of the project?
10. What work has been completed on the project? If you've reported on this project before, what work has been completed since the last report?
11. Will you organize the report by task, by time, or both? Explain.
12. List any tasks attempted that took considerable time but ended up being dead ends.
13. What work still needs to be done on the project?
14. What is your overall assessment of the project at this time?
15. What visuals or attachments might you include in this report?

Design Worksheet

PROGRESS REPORT

Figure 15
Design Worksheet, Progress Report

Instructions: Complete the following prompts thoroughly for your Progress Report.

Beginning
Introduces the subject, provides brief background information, tells the purpose, and prepares the reader for what follows in the report.

Orienting sentence:
Brief summary of project:
Brief summary of main progress to date or since last report:
Purpose statement:

Middle
Presents the details and evidence supporting your conclusions (longest part of the report).

Work completed on project to date (detailed):
Work still to be completed:
Suggested timeline for work to be done:
Potential of project's success:
Visuals or attachments to include:

End
Wraps up the report, and ends on a positive note.

Progress made (brief):
Confidence (or concern) with overall progress:
Sense of cooperation:

Figure 16
Peer Review Worksheet, Progress Report

Peer Review Worksheet

PROGRESS REPORT

Instructions: Read your classmate's report, making comments on the draft. Then complete this worksheet thoroughly. Pretend you are the person receiving the report.

Section A: Read the *Beginning* section and answer these questions.

1. Does the report have a purpose statement? What is it?
2. What is the project the writer is describing?
3. What is the main progress made on the project to date or since the last report?

Section B: Read the *Middle* section and answer these questions.

1. What work has been completed on the project to date or since the last report?
2. What problems with completing the project, if any, does the writer discuss?
3. What work still needs to be done on the project?
4. Does the writer mention a timeline or costs associated with the work still to be completed? If no, should he/she?
5. What conclusions does the writer make about the project's success?
6. Does the *Middle* section provide enough data to convince you that the writer is on top of the project? Explain.
7. Are visuals included in the report? If no, should there be? If yes, does the writer explain and refer to them in the report?

Section C: Read the *End* section and answer these questions.

1. Does the writer briefly state or restate the progress made on the project? Explain.
2. Does the writer show confidence in the overall progress to date or since the last report? Explain.
3. Does the writer tie everything together appropriately and leave you feeling comfortable with the information provided? Explain.
4. Does the writer end with a positive sense of cooperation (that is, offer to provide additional information or meet with the reader)? Does the writer ask for suggestions? Explain.

Section D: Grammar/Spelling/Punctuation

1. Underline all sentences that begin with "There" or "It."
2. Highlight all sentences that are awkward, wordy, or cause you to wonder what they mean.
3. Is the report formatted effectively? Explain.

Section E: Comments

What else can you share with the writer to make this report even more effective?

Communication Strategy Worksheet

Figure 17 Communication Strategy Worksheet, Instructional Document

INSTRUCTIONAL DOCUMENT

Instructions: Complete this Communication Strategy Worksheet for your instructional document. Visualize your reader and answer the prompts/questions thoroughly. The more thorough you are, the easier it will be for you to write the manual.

1. What process or task will you write your instructions on?

2. What does a typical user of your instructions look like?

   - What is the person's age?
   - What is the person's educational background?
   - What is the person's reason for using these instructions?

3. Is your typical user a novice or expert? Place an "X" on the line where your typical user falls:

   Novice ____Expert

4. Describe the information you will include in your instructions.

   - Front matter:
   - Middle matter:
   - End matter:

5. What visuals do you plan to include in the instructions?

6. What else should you consider about the intended user or the content of these instructions (anything not covered earlier)?

Figure 18
Design Worksheet, Instructional Document

Design Worksheet

INSTRUCTIONAL DOCUMENT

Instructions: Complete the following prompts thoroughly for your Instructional Document.

Front Matter

Title page:
Table of contents:
Introduction (audience, purpose, intended use):
Materials needed:
Cautions/Warnings:
Definitions:

Middle Matter

Instructions/Steps/Procedures:
Visuals to include:

End Matter

Conclusion:
Trouble Shooting:
Sense of accomplishment:

Peer Review Worksheet

Figure 19 Peer Review Worksheet, Instructional Document

INSTRUCTIONAL DOCUMENT

Instructions: Read your classmate's manual, making comments on the draft. Then complete this worksheet thoroughly.

Section A: Entire Manual (Read the *entire* document and respond to these questions.)

1. Does the manual have the essential sections (title page, table of contents, definitions, cautions/warnings, introduction, troubleshooting, etc.)? Explain.
2. Does the writer tell you how to use the manual (i.e., is the introduction clear and complete)? Explain.
3. Does the overall design of the manual seem appropriate for the intended user? Explain.
4. Are there plenty of visuals to guide you through carrying out the instructions? Explain.
5. Can you easily follow the step-by-step instructions? Explain.
6. Does each step contain only one instruction? Explain.
7. Does the manual conclude with some kind of final thought? Explain.

Section B: Headings

1. Are the steps chunked under headings? Should additional headings be added? Explain.
2. Do the headings accurately forecast the information following them? Explain.
3. Are headings formatted the same throughout the document (for example, all bolded)? Explain.

Section C: Section Explanations (Major headings should be followed with a brief explanation of what you will be doing and why.)

Are headings followed with a brief explanation of what you will be doing and why? Explain.

Section D: Visuals

1. Do the visuals fit the instructions? Explain.
2. Are the visuals placed appropriately in the instructions? (In other words, are they placed after the step is given?) Explain.
3. Are the visuals an appropriate size (large enough to see the detail, but not too large)? Explain.
4. Are the visuals labeled (for example, *Figure 1 Side View of . . .*)? Explain.
5. Are the visuals documented (for example, *Source: Thomas, p. 30*)? Explain.
6. Are the visuals referred to in the instructions (for example, *See Figure 1.*)? Explain.

Figure 19 Peer Review Worksheet, Instructional Document *(continued)*

Peer Review Worksheet

INSTRUCTIONAL DOCUMENT

Section E: White Space and Page Arrangement

1. Does the arrangement of text and visuals help you understand what to do and how to do it? Explain.
2. Is there too much text on the page? Explain.
3. Do the instructions look hard to follow? Explain.

Section F: Standards (Manuals should present information in the same way consistently throughout; consistency helps the user know what to expect.)

Does the writer consistently present information in the same way? (For example, are all the headings and subheadings phrased the same consistently throughout manual? Do all the steps begin with action verbs consistently throughout? If one caution appears in a box, do all cautions appear in a box consistently throughout?) Explain.

Section G: Grammar/Spelling/Punctuation

Did you find any grammatical, spelling, punctuation, or proofreading errors? If so, mark them on the document draft.

Section H: Comments

What else can you share with the writer to make this manual even more effective?

APPENDIX B

Brief Handbook

Conjunctions and Prepositions

COORDINATING CONJUNCTIONS connect words, phrases, and clauses of equal grammatical rank.

and, but, or, nor

COMMON SUBORDINATING CONJUNCTIONS join dependent clauses to main clauses.

after	even if	so that	whenever
although	even though	than	where
as	for	that	whereas
as if	if	though	wherever
as though	in order that	unless	whether
because	rather than	until	which
before	since	when	while

COMMON CONJUNCTIVE ADVERBS join main clauses to other main clauses in a sentence.

accordingly	however	otherwise	thus
consequently	moreover	so	yet
furthermore	nevertheless	then	
hence	on the contrary	therefore	

COMMON PREPOSITIONS relate a noun or pronoun to some other word in the sentence.

about	before	except	of	toward
above	behind	for	off	under
across	below	from	on	until
after	beside	in	out	up
among	between	inside	over	upon
around	by	into	since	with
as	down	near	through	within
at	during	next	to	without

Nouns, Pronouns, Case

NOUNS

		POSSESSIVE
Singular	doctor, boss	doctor's, boss's
Plural	doctors, bosses	doctors', bosses'

PERSONAL PRONOUNS

Singular	**NOMINATIVE**	**POSSESSIVE**	**OBJECTIVE**
First person	I	my, mine	me
Second person	you	your, yours	you
Third person	he, she, it	his, hers, its	him, her, it
Plural			
First person	we	our, ours	us
Second person	you	your, yours	you
Third person	they	their, theirs	them

RELATIVE PRONOUNS

	NOMINATIVE	**POSSESSIVE**	**OBJECTIVE**
Singular & Plural	who	whose	whom

PRONOUN CLASSIFICATION

Personal pronouns	I, you, he, she, it, we, they
Relative pronouns	who, whose, whom, which, that
Interrogative pronouns	who, which, what, whose, whom
Demonstrative pronouns	this, that, these, those
Indefinite pronouns	one, any, each, anyone, somebody, everyone, each
Reciprocal pronouns	each other, one another
Intensive pronouns	myself, yourself, herself, himself, ourselves, themselves

Verbs

PRINCIPAL PARTS

	Present	Past	Past Participle	Present Participle
Regular	fill	filled	filled	filling
Irregular	lie	lay	lain	lying
Irregular	lay	laid	laid	laying

TENSES

Present	We **fill** all orders promptly.
Past	We **filled** all orders promptly.
Future	We **will fill** all orders promptly.

Perfect Tenses Use Past Participle

Present Perfect	We **have filled** all orders.
Past Perfect	We **had filled** all orders yesterday.
Future Perfect	We **will have filled** all orders by tomorrow.

Progressive Tenses Use Present Participle

Present Progressive	We **are filling** all orders promptly.
Past Progressive	We **were filling** all orders until we . . .
Future Progressive	We **will be filling** all orders until we . . .
Present Perfect Progressive	We **have been filling** all orders . . .
Past Perfect Progressive	We **had been filling** all orders . . .
Future Perfect Progressive	We **will have been filling** all orders for two months by Monday.

Who, Whom, That

Who and **that** refer to people.

Which and **that** refer to places, objects, and animals.

Which introduces nonessential clauses.

That usually introduces essential clauses.

Who is the nominative case.

- Used as subject

Tip: Use *who* when you can substitute he, she, they, I, or we.

e.g., Who is setting up the lab? (He is setting up the lab.)

Whom is the objective case.

- Used as object

Tip: Use *whom* when you can substitute him, her, them, me, or us.

e.g., To whom were you talking? (Were you talking to him?) (You were talking to him.)

Between you and I	Incorrect
Between you and me	Correct

Since *between* is a preposition and *you* and *me* the objects, you must use the objective case *me* and not the subjective case *I*.

Subject-Verb Agreement

A subject must agree in number and person with its verb.

- The *mechanic* is very dependable. (In this sentence, *mechanic* is singular, so it takes the singular verb *is*.)
- These *mechanics* are very dependable. (In this sentence, *mechanics* refers to more than one mechanic, so it takes the plural verb *are*.)

In the sentences above, *mechanic* refers to a person who makes, uses, or repairs machines. Below, you'll find the word *mechanics*, which means something altogether different.

MECHANICS*

1. (used with a singular verb) the branch of physics that deals with the action of forces on bodies and with motion, comprised of kinetics, statics, and kinematics.
2. (used with a singular verb) the theoretical and practical application of this science to machinery, mechanical appliances, etc.
3. (usually used with a plural verb) the technical aspect or working part; mechanism; structure.
4. (usually used with a plural verb) routine or basic methods, procedures, techniques, or details: *the mechanics of running an office; the mechanics of baseball.*

*mechanics. (n.d.). *The American Heritage® Dictionary of the English Language, 4th ed.* Retrieved May 9, 2008, from Dictionary.com: *http://dictionary.reference.com/browse/mechanics.*

Coordination/Subordination

Coordination and *subordination* have to do in a general sense with the relationship between objects. *Coordination* implies equality, elements of equal weight. The relationship in *subordination* is the opposite, that of unequals, one element being more important than or dependent on the other. Thus in business, a person in a subordinate position has less status than her superior, and in grammar, a subordinate construction has less strength than a coordinate construction.

We define the English sentence as a unit composed of a subject (noun)–verb construction (independent clause) that has meaning on its own, i.e., it stands on its own. Thus, "The astronaut lands on the moon" is a sentence that satisfies both criteria—it has a subject (noun)–verb construction (*astronaut lands*) and clear, unencumbered meaning. In a subordinate clause (or dependent clause), the subject (noun)–verb construction is present, but the meaning is not complete. An example of a subordinate clause is "When the astronaut lands on the moon." You see the same subject (noun)–verb construction (*astronaut lands*), but you do not fully understand the meaning; the clause leaves you wondering, "Then what?" Once you finish the thought by adding an independent clause, you get meaning: "When the astronaut lands on the moon, the team at the Space Center will breathe easier."

Choosing which fact to emphasize by making it the independent clause ("independent" grammatically) influences the relationship between two concepts dramatically. Let's look at an example. The reality of the situation described next is this: We are comparing two restaurants, Chez Worley and Maison Fitterling. Chez Worley has earned a four-star rating, has an extensive menu, and is located in Zionsville. Maison Fitterling has also earned a four-star rating, but with a more limited menu, and is located in downtown Indianapolis. Using identical information, look at how the arrangement of the facts changes the emphasis:

Although Chez Worley has a four-star menu, its suburban location argues against it (Don't go to Chez Worley.)

Although its suburban location is out of the way, Chez Worley's extensive four-star menu makes it a clear favorite. (Go to Chez Worley.)

NOTE: Same facts; different emphasis; different interpretation.

More examples:

"Yes, but..." menu is not only variable.

If the menu were the only variable, Chez Worley would win; however, we also need to consider location.

Not only *does Chez Worley have an excellent menu,* **but** *it* **also** *offers the private dining room we need.*

While both Chez Worley and Maison Fitterling are well regarded for their four-star menus, Maison Fitterling is much more centrally located.

Maison Fitterling clearly offers the better location, as it is located downtown, within easy walking distance of our guests' hotels.

Note the concession (Yes, but...); then note reversal.

Punctuation

Punctuation can help the reader understand your message or hinder that understanding. For example, a simple comma can change the meaning of a sentence. Look at these examples:

- The supervisor says the CEO is incompetent.
- The supervisor, says the CEO, is incompetent.

- Add flour, slowly stirring constantly.
- Add flour slowly, stirring constantly.

In the first pair of sentences, sentence one states the supervisor says the CEO is incompetent. In the second sentence, the CEO says the supervisor is incompetent. In the second pair of sentences, the first sentence states the actor is slowly stirring constantly; in the second, the actor is slowly adding the flour.

Punctuation serves an important purpose in our writing. It helps the reader understand how words and ideas relate to one another. Therefore, punctuation is a powerful tool in your repertoire of communication tools. Learning the functions of the different punctuation marks will serve you well as you advance your technical writing style.

COMMA

Of course the comma is a popular punctuation mark (sometimes too popular) and probably the most often used of all the punctuation marks. As already shown, the comma can make or break your attempt to communicate in certain situations. To make it easy to remember how to effectively use the comma, think of it as a separator.

Commas separate in many ways, but here are five common ways:

1. They separate introductory elements (words, phrases, and clauses) from the rest of the sentence.
2. They separate clauses in compound sentences.
3. They separate items in a series.
4. They separate coordinate adjectives.
5. They separate nonrestrictive elements in sentences.

1. **Commas separate introductory elements.** An introductory element can be a word, phrase, or clause.
    - *Finally*, Leif decided to open the door. [word]
    - *Nonetheless*, I completed my recommendation report. [word]
    - *Turning to my right*, I saw the department supervisor! [phrase]
    - When you're finished writing your progress report, I'd like to discuss the next project with you. [dependent clause]
2. **Commas separate clauses in compound sentences.**
    - *A draft of our recommendation report is due next week*, and *the final draft is due the following week*.
    - *I'm feeling a bit overwhelmed with all my responsibilities today*, but *I'll get everything done*.
3. **Commas separate items in a series.**
    - Writing the annual report was *difficult*, *tedious*, *and very time consuming*.
    - The recommendation report includes *background of the problem*, *methodology used to gather the data*, *analysis of the data*, *conclusions drawn from the data*, *and recommendations based on the data*.

    Note: Separate all items in the series with a comma, including the last item.
4. **Commas separate coordinate adjectives.**
    - The *difficult*, *tedious* task of recording the data was taxing Dave's patience.
    - Although peer review is a valuable activity, finding a *competent*, *conscientious*, *available* person to do it can be a challenge.

    Tip: If you can change the order of the adjectives and if you can add an *and* between the adjectives without changing the meaning, they are coordinate and need to be separated with a comma.
5. **Commas separate nonrestrictive elements in sentences.**

    Restrictive means that the word, phrase, or clause limits (or restricts) the meaning of the words modified. In other words, these words, phrases, or clauses are necessary or needed to communicate clearly the intended meaning. Because they are necessary, they *are not* set off with commas. Here's an example:

> Drivers *who do not obey the laws* should lose their licenses.

In this sentence, the words *who do not obey the laws* are necessary to communicate the meaning. We are not saying ALL drivers should lose their licenses. We are saying those "who do not obey the laws" should lose their licenses.

So if *restrictive* means the word, phrase, or clause is necessary, then *nonrestrictive* means the opposite. *Nonrestrictive* means the word, phrase, or clause is *not necessary*, does not limit or restrict the meaning of the words modified.

Because the word, phrase, or clause does not limit or restrict, it *is* set off with commas. Here's an example:

> The two drivers involved in the accident, *who did not obey the law,* should lose their licenses.

In this sentence, the words "who did not obey the law" are unnecessary and serve only as additional information about the two drivers, thus can be omitted without affecting the meaning of the sentence.

SEMICOLON

- Briefly stops the action of the sentence
- Connects items of equal rank
- Is stronger than a comma, but not as strong as a period

1. Use between two closely related sentences.
   - We have no alternative; we must go now.
2. Use between two sentences joined by a coordinating conjunction if the sentences are long and already have commas.
   - The meeting yesterday, the most argumentative and confusing to date, lasted until 7:00 P.M.; and unless something happens to calm people's fears, the next meeting may last even longer.
3. Use between items in a series if any item has commas in it.
   - John Doe, the cardiologist; Jane Smith, the internist; and Susie Jones, the dentist, attended the meeting.

COLON

- Implies a buildup for what follows
- Creates a sense of anticipation

1. Use to introduce a list or a series of items.
   - The following people attended the meeting: . . .
2. Use to connect two ideas or sentences where the second develops, explains, or illustrates the first.
   - He had but one fear: losing the race to his worst enemy. (Use lowercase following the colon if what follows is subordinate.)
   - He had but one fear: He didn't want to lose to his worst enemy. (Use uppercase following the colon if what follows the colon is a complete sentence.)

   Note: A comma or dash could be used instead of the colon in the previous two sentences.)

Don't separate an object from the rest of the sentence with a colon.

- I would like to work in: Indianapolis or St. Louis.

However, when a stacked list follows, you can separate the object with a colon.

- The corporations that manufacture software for this particular problem include:
  ABC Corp.
  Smith & Associates
  XYZ Corp.

PERIOD

1. Use at the end of a statement or command.
   - Yes.
2. Use at the end of a polite request.
   - Will you please handle the production reports for me while I'm away.
3. Use at the end of an indirect question.
   - Kaiton has asked whether an exception can be made to our PTO policy.

QUESTION MARK

1. Use at the end of a direct question.
   - Will you be able to meet with me after 5:00 P.M.?
2. Use at the end of a rhetorical question.
   - Isn't it hard to believe that people would fall for that rumor?
3. Use at the end of a sentence phrased with rising intonation.
   - You actually believe his story?

EXCLAMATION POINT

1. Use with emphatic statements.
   - Congratulations!
2. Use with commands.
   - Come here immediately!

Tip: The exclamation point loses its effect with overuse.

Proofreading Tips

1. Look for these types of common mistakes:
   - Errors in titles and introductory elements
   - Repeated words, especially at the end of one line and the beginning of the next
   - Omissions and substitutions of words, letters, and numbers
   - Reversal of letters, numbers, and words (*two* written as *tow*)
   - Omissions of words at the beginning and ending of a line
   - Words that sound right but are wrong (*your* written as *you're*)
   - Errors in spacing and formatting (indenting some paragraphs but not others)
   - Errors in suffixes (*typed* written as *types*)
   - Errors in closing quotation marks, parentheses, brackets, or commas
   - Problems with appearance (Does the document "look" attractive?)
2. Read each line individually.
3. Read the document backwards.
4. Proofread important or highly technical material more than once.
5. Read numbers digit by digit.
6. Read the document aloud, which allows you to "hear" the mistakes.
7. Follow the text with your finger as you read, the way you did when you first learned to read.
8. Use a sheet of paper to cover all the text below the line you're reading, which helps you focus on only one line at a time.
9. If the material is complex or involves many numbers, have someone help you proofread. One person reads aloud the original material, noting all punctuation and any important details of style and format (e.g., paragraphing and capitalization), while the second person proofreads the copy.

CAUTION: Do NOT rely solely on your computer spell and grammar checkers. Proofread and edit everything carefully. Always have a good dictionary and other resources available.

Index

M

N

O

P

Q

R

S

T

U